636.3   Simmons, Paula.
SIM
         Raising sheep the
         modern way.

$9.95

DATE			

Raising Sheep
The Modern Way

UPDATED & REVISED EDITION

PAULA SIMMONS

A GARDEN WAY PUBLISHING CLASSIC

STOREY COMMUNICATIONS, INC.
POWNAL, VT 05261

Cover by Ken Braren
Book designed by Cindy McFarland
Typesetting by Accura Type & Design
Illustrations on pages 202 through 211 by Elayne Sears, and illustrations on pages 63, 66, 90, 97, 123, 248 by Brigita Fuhrmann

Copyright © 1976 by Storey Communications, Inc.
Copyright © 1989 by Storey Communications, Inc. Revised edition

Printed in the United States by Courier
Twelfth printing, Revised edition, November 1993

Garden Way Publishing was founded in 1973 as part of the Garden Way Incorporated Group of Companies, dedicated to bringing gardening information and equipment to as many people as possible. Today the name "Garden Way Publishing" is licensed to Storey Communications, Inc., in Pownal, Vermont. For a complete list of Garden Way Publishing titles call 1-800-827-8673. Garden Way Incorporated manufactures products in Troy, New York, under the TROY-BILT® brand including garden tillers, chipper/shredders, mulching mowers, sicklebar mowers, and tractors. For information on any Garden Way Incorporated product, please call 1-800-345-4454.

Library of Congress Cataloging-in-Publication Data

Simmons, Paula.
    Raising sheep the modern way / Paula Simmons. — Updated
and rev. ed. — Pownal, VT : Storey Communications, c1989.
      p.   cm. — (A Garden Way publishing classic)
    ISBN 0-88266-529-4
    1. Sheep.   I. Title.
SF375.S56   1989
636.3—dc20
                                  88-45488
                                      CIP

Contents

Modern sheepraising has shown a real trend toward the small situation, with emphasis on profitable self-sufficiency. With more and more people keeping a few sheep, the average number of sheep on small farms in this country is similar to the size of the average farm flock in Switzerland—seven sheep.

This small number per flock makes it more urgent that there be no losses due to disease or neglect, and this requires greater knowledge of the fundamentals of sheep health, and of preventive care, and of proper medical treatment, should that become necessary.

While the fundamentals are the same, this new edition includes many important new developments, with a totally updated medical chapter featuring some remarkable vaccines that now offer protection against some of the diseases that threatened sheep in the past. There is also a new chapter on guardian dogs, a natural answer to the predator problem.

Dedication

Thanks to Bill Hess of Sheepman Supply and to Bill Schaefer, D.V.M., for their helpful advice.

Above all I must acknowledge a large debt of gratitude to my veterinarian friend and adviser Darrell Salsbury, D.V.M., for generous technical and medical guidance. He wrote me the following poem just for this edition:

THE SHEPHERD'S LAMENT

Now I lay me down to sleep,
Exhausted by those dog-gone sheep;

My only wish is that I might
Cause them not to lamb at night.

I wouldn't mind the occasional ewe,
But lately it's more than just a few:

Back into bed, then up again,
At two o'clock and four A.M....

They grunt and groan with noses high,
And in between, a mournful sigh,

We stand there watching nature work,
Hoping there won't be a quirk:

A leg turned back, or even worse,
A lamb that's coming in reverse.

But once they've lambed we're glad to see
That their efforts didn't end in tragedy.

There's no emotion so sublime
As a ewe and lamb that's doing fine.

I'm often asked why I raise sheep,
With all the work and loss of sleep;

The gratification gained at three A.M.,
From the birth of another baby lamb—

How can you explain, or even show?
'Cause only a shepherd will ever know!
            D. L. SALISBURY, D.V.M.
                    APRIL 1, 1988

CHAPTER 1

Starting With Sheep

IT IS PREFERABLE to "grow" rather than "buy" into sheep. Starting small gives you the opportunity to get low-cost experience. If you start with fewer sheep than your pasture will support, you will be able to keep your best ewe lamb each year, for a few years at least. And after a few years, as any of your purchased ewes reach an unproductive age, they can be replaced by keeping one of your best lambs.

If you haven't any preference of breed, consider the predominant one in your area. It is likely to be well suited to the climate, and buying close to home saves shipping costs and a stressful ride for the animal. You can also get replacement rams more easily, even trading with other breeders nearby, when you have used yours for a while and want to avoid inbreeding.

PUREBREDS

A purebred possesses the distinct characteristics of a specific breed, and either is registered or eligible for registry in the association of that breed.

The advantages in purebreds are greater uniformity in appearance and production, and a chance of income from the sale of breeding stock. In theory, they may be in better health, as the owner probably would take better care of a valuable animal.

The disadvantages are higher initial cost, plus the cost of registering each lamb, with no better price for either the meat or the wool than if they were not registered. Also, the financial loss is greater if one dies.

Sheep that have the characteristics of a particular breed, without the registration papers or the assurance of their breeding, are called *grades*.

THE RAM

Often it is a good investment to buy a purebred ram to use on grade ewes, for his good characteristics will show up in every lamb that is born. (More on this in Chapter 6.) Many experts feel that the ram's breed exerts more influence on the lambs than does the ewe's. This may be true for most factors, but does not pertain to twinning, since the ewe is the one that drops the eggs and makes twinning possible. The important influence of a twin ram is on its *daughter*, who will have a significantly higher chance of shedding two eggs at ovulation than will a daughter of a ram with a background of single births. When buying a ram, he should have a negative ELISA test and be free of hoof rot.

WESTERN AND NATIVE EWES

In predominately sheep-raising parts of the country, some sheep are classified as "native" sheep, and some as "western" or "range" sheep. The ones called native are mostly meat-type animals, large-sized, prolific and usually black-faced. The western sheep are usually fine-wool sheep, or a cross of fine-wool and long-wool breeds. The fine-wool sheep were often preferred on the western range, not for their wool, but for their superior herding instinct.

Purebred rams are almost always used on western and native ewes, the breed of ram being determined by the specific market, such as a wool market or a lamb market.

BUYING

It is not always possible to heed the following criteria, because your selection may be limited. But if you know what is undesirable, you can better evaluate the price being asked for the sheep.

Try to avoid the following:

1. Any condition resembling pink eye, or any eye damage.
2. Teeth missing. A sheep with missing teeth can't eat well and may require special care and feeding.
3. Lower jaw not matching upper jaw properly, either overshot or undershot.
4. Lumps in the udder. These may indicate mastitis, so that a lamb would require supplemental bottles, or be a complete bottle baby.
5. Limping sheep. This may indicate hoof disease.
6. Untrimmed feet, turned up at the toes like skis, or overgrown and turned under at the sides.
7. "Bottle jaw." Lumps or swelling under the chin, usually caused by severe internal parasite infestation.
8. Extremely thin ewe. Unless she has just raised twins or triplets, she may have a disease (such as OPP), or a heavy load of parasites.
9. Extremely fat ewe. She may not breed, or if bred she may have trouble lambing.
10. Wool going too far down on legs. It is more trouble to shear.
11. Wool covering face. Shearing the face takes time. Wool blindness inhibits eating and mothering. In addition, Australian tests have proven muffle-faced ewes to be less fertile and productive.
12. Fine-boned sheep. Those with larger bones are more hardy and productive.
13. Small size. Potentially smaller production of both lambs and wool. Those that are undersized for their particular breed are not recommended.
14. Narrow or shallow-bodied. Their lambs lack good carcass conformation.

Open-faced sheep (left) has been found to be more fertile and productive than muffle-faced ewe (right).

15. Old sheep. Those over seven or eight years old are progressively less profitable. Even if they have good teeth, they are a poor investment *unless* their price is low.
16. Ragged, unattractive fleece. This may indicate sheep is scratching itself because of ticks or mites. Part the fleece in several places and look closely.
17. Sheep with runny droppings. May be caused by lush seasonal pasture, but also could be caused by internal parasites. Ask how recently they were wormed, and what drug was used.
18. Sheep that were single births. Twinning is quite hereditary, primarily influenced by the ewe and more profitable.

IMPORTANCE OF TWINS

Although multiple births do require more attention and care, the profits seem well worth the effort. A 1987 University of Wisconsin analysis stated that it would require 5,721 ewes producing one lamb each to generate a $25,000 profit, and only 353 ewes producing two lambs each to equal it. Those figures do sound strange, but consider the vast reduction in the amount of grain and hay expenses for the smaller number of ewes, to produce double the number of lambs.

SHEEP AGE VS PRICE

The age of the sheep is important in relation to the *asking price*. Just how many more fleeces and lambs can this ewe be expected to produce? If she is quite old, how much additional feed will she need to compensate for her poor teeth? Again, there is the importance of multiple lambs. Does this old ewe have a history of twins or triplets? If so, this makes her more valuable, assuming she gets the care that makes more lambings possible.

TEETH

You can tell a sheep's age by its teeth, but only up to a certain age. As shown in these photographs, a lamb has eight small incisor teeth until it reaches approximately one year of age. Each year thereafter, one pair of lamb teeth is replaced by two permanent teeth that are noticeably larger. By the time a sheep is four years old, all the lamb teeth have been replaced with permanent teeth, and it is no longer possible to tell its age *accurately* by the teeth. You can only estimate by the condition of the teeth. (In case you wondered, sheep also have twenty-four molars.)

Teeth wear drastically shortens a sheep's life. The incisors are all on the bottom jaw, and as the teeth wear down, the amount of tooth below the gumline (about 1/2

Teeth are an indication of the age of sheep.

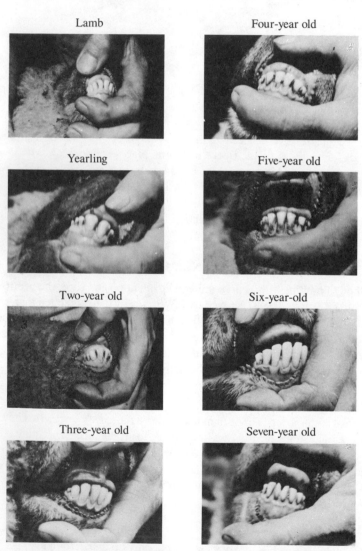

Lamb	Four-year old
Yearling	Five-year old
Two-year old	Six-year-old
Three-year old	Seven-year old

(Michigan State University Cooperative Extension Service)

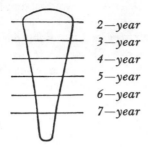

2—year
3—year
4—year
5—year
6—year
7—year

Approximate annual wear of sheep's teeth.

inch) is gradually pushed out to help compensate for the wear. This is partly why the teeth of an old ewe look so much narrower.

The wider part at the top of the tooth is being worn back toward the narrower center-part of the tooth. The tooth is also being pushed up a bit, and the pushed-up part is from below the gumline, and narrower. The gaps between the teeth reduce the efficiency of the ewe's bite. If you listen to an old ewe grazing, you will hear a squeaking of grass as it slips between her narrowed teeth.

On very low or overstocked pasture, teeth wear down faster from soil and sand. The closer to the soil sheep graze, the more dirt and sand they ingest with their food, and these wear down their teeth like sandpaper. On short pasture, ewes also must take more bites to get every pound of grass they eat, and this wears on the teeth.

SHEEP: ACCORDING TO THEIR TEETH

Solid Mouth. Having all adult teeth in place (up to about four years of age).

Spreaders. These are older. Teeth show wear with the under-gum portions, which are narrower, moving up into position.

Broken Mouth. These have some teeth missing. You may get one or two seasons of lambs from them.

Gummers. Sheep that have lost all of their front teeth. A very poor buy.

I would add that gummers may do better than a badly broken mouth sheep, as the gums harden so that they can still chomp off grass, while broken mouths with two or three teeth cannot get a good grip on the grass. If you have an old ewe with a broken mouth who is down to one or two loose teeth, and you are determined to keep her, those few teeth should be pulled with pliers, as they prevent the lower gum form making contact with the upper pad. When the gum hardens, she can eat grass more efficiently than with only a few teeth. She will still need special attention in feeding.

British magazines have reported an orthodontist who has designed dentures for sheep. And in France, Dr. Gilles Raoult, a dental mechanic, has made false teeth for both cows and sheep, charging about $40 to $45 per animal. These extend the life of a ewe by at least five productive years, which would be worth it for a twinning ewe with a good fleece.

If you have a dental-student friend who wants to make dentures for your old ewe, have him or her look first at the teeth of a young ewe. Notice that these front teeth are gouge-shaped, concave without and convex within. This is one reason why sheep can crop the grass closer to the ground than other farm animals. (Another reason is the narrow, flexible muzzle, which is divided by a vertical cleft.)

TOOTH GRINDING ON OLDER EWES

A process of grinding down the teeth of mature sheep has been gaining in popularity in Australia. The Australian device used consists of a 45-cm gag made of steel tubing with two guides to hold the sheep's mouth open, a half-circle plate between the guides to hold the tongue out of the way, and a hand-protector at each end. This mouth piece arrangement gives the access for the 100-mm disc which is used for the once-in-a-lifetime grinding (usually at about 4 1/2 years of age) of the teeth of valuable productive ewes. This is claimed to give immediate and long term improvements, such as weight gains within the first month, increase in wool clip, loose teeth setting solid again, and healthier animals because of their improved foraging ability.

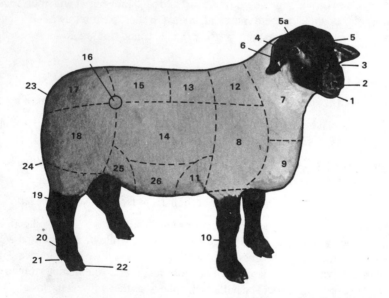

(1) mouth. (2) nostril. (3) face. (4) eye. (5) forehead. (5a) poll. (6) ear. (7) neck. (8) shoulder. (9) brisket. (10) foreleg. (11) foreflank. (12) top of shoulders. (13) back. (14) paunch. (15) loin. (16) point of hip. (17) rump. (18) thigh. (19) rear leg. (20) pastern. (21) dew claw. (22) foot. (23) dock. (24) twist. (25) rear flank. (26) belly. Numbers 18 through 24 represent the leg of mutton. (Suffolk Sheep Association)

BUYING OLDSTERS

You can get started in sheep raising with the least outlay by purchasing old ewes, someone else's culls. Frequently they may once have been the previous owner's best ewes. Their years are numbered, and hence the initial cost will be low. If you keep the very best of their lambs, you will be in business.

A commercial grower will consider a ewe to be old at seven or eight years of age, while with good feed, her lambing capability goes on to ten or twelve years. She will do better for you than for her former owner, *if* she does not have to compete with younger ewes for food. So, you will be doing the sheep a favor, and her owner will probably be happy to find a sale for her.

PREGNANCY OF OLDSTERS

Don't *over*-feed them in the early months of pregnancy, as they will still need to have better feed in the latter part of pregnancy to avoid toxemia. Encourage them to exercise, since that contributes to their good health. Their lambs may need at least supplemental bottle feeding (with lamb milk-replacer, see details in Chapter 10) once or twice a day, if the ewes' milk supply is not ample. Start the bottle feeding (very small amounts) the day the lambs are born, to ensure that they will continue to accept the bottle. Since you are only their extra mama, it is not as confining a chore as if they were real bottle lambs, with no sheep mama.

AVOIDING SHIPPING FEVER

To prevent shipping fever, it is helpful to give a shot of Combiotic before transporting the sheep. Move them in mild weather, if possible, and avoid rough handling and overcrowding in transporting. If a large number of sheep will be moved an antibiotic can be added to their water ahead of time.

AVOID PROBLEMS

Feed the same type of grain at your farm as the sheep you bought were accustomed to eating. Ask the owner what kind; if it is not easily available, buy some from the owner. Then gradually change the sheep from their accustomed grain to whatever you intend feeding. Never change abruptly.

To avoid scours or bloat, sheep should be given their fill of dry hay before being turned out on a pasture more lush than they had before.

Ask the owner when the sheep had their feet trimmed last, and see if he or she will trim one of them while you watch. This is an easy way to learn how it is done, and how the feet should look when properly trimmed.

You should know how recently the sheep were wormed, and what drug was used. If they are due for a worming, maybe the owner will do it while you watch.

It is important to find out what vaccinations have been given to sheep you are purchasing. You should know what disease problems may have troubled the flock from which they come.

It is very good insurance to require that all sheep have tested negative for OPP (Ovine Progressive Pneumonia), be free of hoof rot, and that a ram has tested negative for Ram Epididymitis.

CHAPTER 2

Sheep Breeds

ANY DISCUSSION of sheep raising begins with mention of sheep breeds, whether you are purchasing your first sheep or adding to an existing flock. The selection of the "right" breed may seem momentous, but it may not be practical to bring in an exotic breed from a far place, or it may just not be affordable. The best route may be to use a more available breed, knowing that a careful and patient breeding program could upgrade any breed, and may even give many of the desired qualities of the less-available one. Just the addition of one special ram, at a later date, might accomplish your breeding goal.

Different breeds were developed in response to market needs and conditions under which they had to be raised. However good the particular breed sounds, it may not be the best choice for your situation. Such things as climate, pasture, and how much time and money can be invested in their care all have to be considered.

Some breeds have a higher incidence of multiple lambing, which is fine if you are able to give them sufficient attention to ensure survival and good growth. Twins and triplets, without supplemental grain feeding, will grow more slowly than singles.

There are breeds that can be managed to lamb more than once a year, but out-of-season lambing does not usually correspond to the best time for pasture grazing, so both ewes and lambs will need special feeding.

More recent breeding programs have placed less emphasis on visual appearance, and more on carefully measured productive characteristics, such as food conversion and weight gains, fast growth of lambs, mothering ability, prolificacy, and clean weight of wool.

The development of any new breed is slow, even with modern technology—which includes frozen embryos, frozen semen, ova transfers, and computerized record keeping combined with vigorous selection and extensive culling. Within the past ten years, several fairly new breeds have begun to make news. The most prominent are the Booroola Merino, the Cormo and the California Red.

Some older breeds, either rare or neglected, are now getting attention for specialized purposes. These include the Jacob, which was once deemed an "endangered species," and the Shetland sheep with a newly established large flock in Vermont. The Romanov is already being released from quarantine in Canada, and the Perendale is rapidly gaining favor with spinners.

A chart at the end of this chapter was prepared by the *National Livestock Producer*. Seven sheep experts across the country rated the major sheep breeds for twelve different traits, and then gave them a final breed ranking by tallying them up on a point system. A few traits were missing, and another chart has been prepared by Robert M. Jordan of the University of Minnesota, which lists some of these other traits, with his evaluation.

The main value of these charts is definitely not to see which breed is in first place or second place, but to show a fairly unbiased listing of the strong points and weaknesses of each. This makes possible a more informed choice of which breed is suitable for your own needs.

In a situation where it is not possible for someone to be at home during the day in lambing season, then "ease of lambing" would be more important than "growth rate." And if you live in a climate of extremely hot summers, then "heat resistance" would be more urgent than "longevity." You will have different needs, if you intend to spin your wool or sell to spinners, than will the person who is selling great quantities to a wool dealer, or who is primarily interested in selling locker lambs.

In evaluating breed characteristics, keep in mind that wool on the sheep's legs should be considered a disadvantage because it is unusable and makes shearing more time-consuming. Wool on the face, which is very inheritable, is another disadvantage. Tests have shown that an open-faced ewe (with no wool on the face) will raise more lambs and heavier lambs than wooly-faced ewes. Also, open-faced ewes do not suffer wool blindness or collect burrs on the face.

Skin folds, in general, are also undesirable. They do produce a higher grease weight of fleece, but they also mean higher shrinkage. For handspinners, the excessively greasy fleeces are harder to wash. Folds make shearing more tedious and cause more second cuts, and, since maggots can hatch and thrive in moisture-retaining folds, the folds predispose sheep to fly strike. Skin folds usually indicate a somewhat lower fertility and productivity, according to the U.S. Sheep Industry Development program. Folds are inheritable, and found mainly in some of the fine-wool breeds.

This chapter is intended to give a brief summary of the qualities of some of the major sheep breeds.

BARBADOS BLACKBELLY

The Barbados is a dark tropical hair sheep, originally from the Island of Barbados in the West Indies and said to have developed from West African stock. Some of the recent interest centered around them comes from their trait of lambing almost twice a year. They are prolific, hardy, and breed out of season. The ewes are nervous around strangers, protective of their young, and need high fences if worked much. Young ewes breed prior to one year old, usually at five to seven months of age, and are very good mothers.

The breed is reputed to be resistant to internal parasites, and field trails have confirmed this, showing 236 parasite eggs/gm/feces compared to 2,300 or 2,077 or 1,490 of other breeds and crosses (Lemuel Goode, North Carolina).

Blackbelly ram and ewes. (Dr. Lemuel Goode, North Carolina Experiment Station)

The North Carolina Experiment Station has been crossbreeding since 1971, using Barbados x Dorset as well as Dorset x Landrace, to improve overall productivity of ewe flocks by raised reproductive performances, plus resistance to heat stress. Some areas of North Carolina have serious problems caused by heat, including ram infertility, failure of ewes to exhibit estrus, fertilization failure, early embryo death, and impaired fetal development.

Projects are also going on in crossbreeding for year-round lambing ability, the best documented one being by Glenn Spurlock of the University of California. There the average lambing interval has been from 6 1/2 to 7 months, although in multiple lambings (three or four lambs) this interval is about eight months, multiple lambing appearing to retard the return to breeding.

Between 250,000 and 500,000 of the Barbados are now in Texas, many being crossed with Rambouillet, and some crossed with the European Moufflon which is a wild sheep, and used for hunting on game ranches. Since the Barbados withstands both heat and cold very well, it is especially suited for Texas.

Some of the Barbados in this country have horns, usually a result of some prior crossing. A cross with Dorsets improves wool, decreases nervousness, and increases docility. There is some crossing done with handspinning wool in mind, obtaining the dark fleece of the "Barbs" and their accelerated lambing, but crossed to a sheep of a better spinning-quality wool. The crosses then show even more variety of dark coloration than the pure Barb.

Booroola Merino. (Agriculture Canada)

BOOROOLA MERINO

This strain of Merinos started with a single prolific ewe at the Booroola farm in Australia, and has developed into a breed noted both for high quality long-staple fine wool and for its amazing prolificacy. The breed's multiple lambing results from a single major gene that affects ovulation, unlike the Finnsheep in which ovulation is controlled by a large number of genes. Australia's CSIRO* donated three rams to the New Zealand Ministry of Agriculture in 1973, and both countries have carried on extensive research. By 1981, Booroolas were being exported to Brazil, Chile, Hungary, Uruguay, England, Canada, Scotland, and the United States.

The Booroola gene being dominant, one dose is sufficient to produce a large effect on ovulation rate and hence on lambs born. While lambing percentages can be increased to equal those of breeds such as the Finnsheep, there need be no increase in undesirable characteristics, often associated with the other prolific breeds.

Research is now being conducted at the U.S. Meat Animal Research Center in Nebraska, where Booroolas will be compared with Finnsheep for reproductive traits as well as for survival, growth and carcass characteristics, and their usefulness in crossbreeding.

Rams, ewes, frozen embryos, and frozen semen are available for export from the Booroola Programme, Haldon Station, Fairlie, New Zealand, as well as from

---

*Commonwealth Scientific Investigations and Research Organization

Australia (subject to normal importation regulations). They are more easily available in Canada and the United States, from the following addresses:

Breeding stock

J. Sloan Booroolas
10 Campbell Cr.
Willowdale, Ont. M2P 1P2
Canada

Breeding stock and semen

Booroola—Nevada
Box 159
Galconda, NV 89414

CALIFORNIA RED

In the early 1970s, Dr. Glen Spurlock of the University of California at Davis began crossing Tunis and Barbados sheep in order to establish a new strain with superior qualities for the production of both wool and meat. The result has been the rapid development of flocks that are now quite consistent in maintaining the breed traits and characteristics.

California Reds are a medium-sized dual-purpose sheep. Rams weigh from 200 to 250 pounds, and ewes 130 to 160 pounds at maturity, and yield fleeces of about 7 to 8 pounds annually. The texture of the wool is silky and contains reddish hair, which makes it highly desirable to spinners and weavers. Rams are active and aggressive even in the hottest weather, and the ewes are good milking mothers, free of lambing problems. Most breeders plan three crops of lambs in a 24-month period. Lambs are red at birth but the wool eventually becomes beige at maturity.

California Red, four-year old ram. Raised by Paulette Soulier.

Cheviot ewes. Breeder is Virginia Rowell in California.

CHEVIOTS

There are two distinct types of Cheviots. The smaller Border or Southern Cheviot was improved by selection from the original stock rather than by crossbreeding, and is the predominant type of Cheviot in this country.

The Cheviot started as a mountain breed, native to the Cheviot hills between Scotland and England. It is extremely hardy and can withstand hard winters, and graze well over hilly pasture and in high altitudes. It lacks the herding instinct needed for raising on open range, but does well in a farm flock. On scant pasture, the Cheviots spread out and get all the available feed.

They are active and high strung, being alert both in appearance (erect pointed ears) and behavior. They are good mothers despite their nervousness, and their newborn lambs are more hardy than many other breeds. Because of their small head, they experience few lambing difficulties, and they raise a good meat lamb. They are short and blocky, with clean faces, strong noses, and black nostrils and lips that combine with the sharp ears to make them a very recognizable breed. They have a lightweight fleece, medium wool about 48s to 56s that is easy to use in handspinning. The "other" Cheviot is the North Country Cheviot, a larger size animal, more of a Scottish breed, and has a more pronounced Roman nose. Its size is the result of earlier crossbreeding.

CLUN FOREST

This is a new breed in this country, and not well known yet. The first six ewes were brought from Northern Ireland to New York in 1959, but the first large importation was not until 1970, from Shropshire, England. Because of strict quarantine regulations, no breeding stock from this shipment could be sold until 1973.

Clun Forest ewes are prolific, almost always having twins. With narrow sleek heads and wide pelvic structure, they show unusual ease of lambing without assis-

Clun Forest ewes, each with twins under one week old. Breeder is Mrs. Warren G. Menhennett, Cochranville, PA.

tance. With the forty ewes imported in 1970, assistance was only given to one ewe in three years of lambing. The owner, Mr. Turner, afterward said that it might not have been necessary (she was having triplets). They have a strong mothering instinct, even when lambing as yearlings. An average of 80 percent of Clun ewe lambs will breed as lambs at eight or nine months old, and lamb as yearlings.

Geoffrey Bowen, in Wool Away, claims that Clun Forest twin lambs easily attain 45 pounds dressed weight at four months of age, showing the fast growth of these lambs, attributed to the richness of the ewe's milk—not more fat content than other sheep, but more protein. They are adaptable to all climates, being successfully grown from the rainy areas of England to the hot dry climate of Ghana, in low, marshy places of England to high hills in Wales and Scotland. They are good foragers.

Another quality that makes them valuable is their longevity, bearing twin lambs and good fleeces to age ten and twelve. Their medium wool is of 58s count.

The Clun Forest Breeders Association members have voted unanimously to sell only rams for breeding purposes that have been born a twin, triplet, or better.

COLUMBIA

The Columbia is an American breed, developed since 1912. Started in Wyoming, it was sent to the U.S. Sheep Experimental Station in DuBois, Idaho, in 1917. It is the result of a Lincoln ram and Rambouillet ewe cross, with interbreeding of the

Columbia ewe. This is the 1974 national champion.

resulting crossbred lambs and their descendants without backcrossing to either parent stock. The object of the cross was to produce more pounds of wool and more pounds of lamb. Their large robust frame and herding instinct have made them excellent for western range purposes, but they have also proved admirably adaptable to the lush grasses of small farms in all parts of the country. Heavy wool clip, hardy and fast growing lambs, open faces, and ease of handling are characteristics appreciated everywhere.

They have medium wool, in the 50s to 60s range, but predominately about 56s. It has light shrinkage, and is an excellent fleece for handspinning.

It is an all-white breed, polled, and open faced. Since it is such a relatively new breed, even more improvements are expected, and there are Columbias being raised at a number of experiment stations, including the Washington State University Research Center at Prosser.

COOPWORTH

The Coopworth is a new breed in New Zealand, a cross of Border Leicester x Romney, and has unusually strict registration requirements. Performance recording is mandatory. For "merit" registration, a ewe must have raised six lambs by the age of 4 1/2 years, and be above average for yearling fleece weight and total weaning weight of her lambs. If she fails to rear two lambs on more than one occasion after qualifying for the merit register, she loses the merit status.

A ram must be of multiple birth, and born within the first twenty-one days of

Cormo breed. (American Cormo Sheep Association)

lambing. He must be out of a merit ewe or one with a lifetime 170 percent lambing record. His sire has to be at least 2.2 pounds above the average of his contemporaries for adjusted weaning weight* and 0.22 pounds above average for yearling fleece weight.

CORMO

This breed was developed in Tasmania, guided by principles originated by Dr. Helen Newton Turner, believed to be the world's leading sheep geneticist. A group of Australian scientists selected Tasmanian stud Corriedale rams to cross with superfine Saxon Merino ewes, resulting in the Cormo breed whose background is one-fourth Lincoln, one-fourth Australian Merino, and one-half Saxon Merino. The Cormo's outstanding qualities are the fine, well-crimped wool, excellent conformation, fast growth, high fertility, and the ability to thrive where there is heavy snowfall, severe climatic conditions, and rough terrain. Ewes have an average fleece of 11 pounds, yielding 70 to 73 percent clean weight, with a fleece uniformity (90 percent within 2 microns or the average) that makes it valuable to industry. Handspinners find it the most exciting of the finewool breeds.

No pedigrees are kept. Sheep are numbered and computer management makes the Cormo the most strictly scientific genetic improvement scheme in sheep history.

---

* All lambs reared as singles are excluded when establishing the adjusted weaning and fleece weight.

Corriedale ewe and twins. (*Shepherd* magazine)

Cotswold ewe. (*Sheep Breeder and Sheepman* magazine)

CORRIEDALE

The Corriedale is a Merino-Lincoln cross, developed in Australia and New Zealand and first brought to this country (Wyoming) in 1914.

Its dense fleece is medium fine, 56s grade, with good length and softness and light shrinkage, somewhat between medium wool and long wool, a favorite of handspinners in many areas of the country. Its face is clean of wool below the eyes, and the sheep is hornless. Bred as a dual-purpose sheep, it has good wool and good meat for greater profits, and is noted for a long productive life, which means greater return on your investment. Because of a marked herding instinct, it is also a good range breed.

COTSWOLD

The Cotswold has very long coarse wool, 8 to 12 inches or more, that is wavy, hanging in pronounced ringlets, with wool hanging over the forehead. There is very little shrinkage in this fleece, but it is sometimes excessively hairy on the thighs.

The Cotswold thrives on dry rolling land, and is similar to the Lincoln and Leicester. It is a very large sheep, whose most characteristic feature is the long tuft of wool hanging over the face.

DEBOUILLET

Development of the Debouillet was begun by Amos Dee Jones in New Mexico in the 1920s. This was a cross of Ohio Delaine Merino rams and Rambouillet ewes, the successful crosses showing the length of staple and character of the Delaine fleece and the large body of the Rambouillet. By 1927 the ideal type was attained, and a line breeding program begun. The breed was registered in 1954, starting with 231 rams and 1,587 ewes.

Debouillets are open faced below the eyes and over the nose, have a good belly wool covering, and shear a heavy fleece of long staple fine wool. Rams can be horned or polled. Even under adverse conditions, the ewes produce desirable type market lambs of excellent weight.

Debouillet lambs that are eligible for registration by bloodline must be one year of age and in full fleece when inspected by an association inspector. Wool must by 64s grade or finer, with staple of 3-inch minimum, and deep close crimp.

DORSET

The Dorset Horn ewes and rams both have horns. The Polled Dorset are bred to be without them, on both sexes. The first Polled Dorsets were developed at North Carolina State College and first registered in 1956.

Dorsets originated in England, but their history is not well known, although it is believed that the breed developed more by selection than by crossbreeding.

The Dorset has very little wool on the face and legs and belly, and a lightweight fleece that is good for handspinning.

Dorset ewe with lamb. (Continental Dorset Club)

This sheep has a large coarse frame, with white hooves and pink skin. The ewes are prolific, often twinning. They are good milkers, having even been kept in dairies at one time in England, so their lambs grow well. They are good mothers, and have a special feature of early breeding for fall lambing, and it is even possible to lamb twice a year. A *Shepherds Guide* of 1749 described Dorset Horn sheep as "being especially more careful of their young than any other."

A Dorset ewe has a good appetite and good digestion, and can take a lot of feed and feed her lambs well to get them to market by the time many other breeds are just lambing. These out-of-season lambs can command good prices.

DRYSDALE

This is the New Zealand Romney with the "carpet wool gene," a genetic mutation that has happened in some Romney sheep there. It was first discovered by Dr. F.W. Dry of Massey University in the 1930s. A lamb with that N gene can be identified by a patch of pure hair that is noticeable under the shoulder at birth. This was a very uncommon occurrence, and by considerable genetic research, breeding and selection, was finally developed into an established breed.

Flocks have remained closed since 1967, with a ban on exportation. Overseas Carpets Ltd. distributes and controls the rams, to protect the valuable output of their use, and farmers raise them under contract.

The wool produced is a hairy 28s to 36s count, with a 15-inch growth in eleven months. It grows so fast that some are sheared every eight months. This uniform grade of carpet wool is valued for the manufacture of resilient carpets with long life, and without undue "tracking" of the surface.

EAST FRIESIAN

This is a German breed, raised primarily for milk, particularly for cheese production. It is the highest milk producer of all the European breeds, as well as being very prolific. While it is a large-sized sheep and lambs have a good growth rate, it is not considered an especially good meat carcass.

It shears a heavy fleece of 48s to 50s wool.

In crossbreeding for increased milk production to accompany prolificacy, this would be a valuable genetic addition. In southeastern France, this is one of the three milk breeds crossed together for the famous supermilking sheep that are the basis of the Roquefort cheese production there.

FINNSHEEP

The Finnish Landrace is native to Finland, fairly new in this country, but a fast-growing breed. Its main value is for crossbreeding to introduce its unusual prolificacy, being called the sheep that "lambs in litters." Up to 1/8 Finnsheep in a cross will increase lambing percentages somewhat, although the same thing can be done with good culling and breeding for twins and triplets in your own stock.

When these sheep have quadruplets and quintuplets and sextuplets, it is customary to leave two or three on the ewe, and feed the rest on lamb milk-replacer, after they get the colostrum. They are good mothers, and topped all the breeds in the "breed chart" for ease of lambing.

First brought to the United States about 1966, they have expanded rapidly, and more than 2,600 were registered as Purebred Finnsheep by 1975, with many more not registered. This breed will register black sheep, and has a fairly high incidence of black lambs, for this feature has not been purposely eliminated in the breeding.

The ewes will breed young, most lambing before they are a year old. Ewes that are to be bred at the age of six to seven months should weigh at least 100 pounds before breeding, and at least 125 pounds at lambing. They will need special feed and care

Finnsheep ewe and quintuplets. This remarkable ewe, Langelmaki 57A, had a total of twenty-seven lambs in seven years, including these quintuplets which she raised without assistance. Her owner is Werner Grusshaber of Finnlamb Farm, Brantford, Ontario, CANADA.

This Finnsheep Cross ewe is half Dorset, half Finn Cross. She raised these quadruplets without help, and she alternates between triplets and quadruplets, showing the success of the half Finn Cross. (Mary Sue Ubben, Breinigsville, PA)

during gestation, to meet the nutritional needs of their multiple lambs. If these young ewes are to grow and reproduce at the same time, the breeder must give extra care.

Finnsheep have naturally rather short tails, which don't always require docking.

FINNSHEEP CROSSES

There is widespread use now of Finnsheep for crossing with other breeds to increase their lambing percentages. In commercial flocks, the meat carcass of the quarter Finn crossbred lambs appears to be indistinguishable from that of the meat breed used in the cross, while greatly improving lambing percentages.

HAMPSHIRE

The Hampshire is one of the largest of the medium wool meat-type sheep with very rapid growing lambs. While they do not do well on rough or scanty pasture because of their size and weight, they do nicely on good pasture and the lambs can usually be marketed off grass. The ewes are good milkers and fairly prolific, but do not always lamb easily, probably due to the large head and shoulders of the lambs, which are quite heavy at birth.

The Hampshire has a large head and ears, is hornless, and the dark face is closer to a rich dark brown than to black, with legs of the same color. Its fleece is light-weight, medium wool (48s to 56s), and fairly short.

The rams are one of the favorite breeds to use on fine wool range ewes, to attain the fast growth of market lambs. The rams are also noted for their keen sense of smell, important in detecting estrus in the ewes.

Lively Hampshires head for the pasture. (John D. Wibbels, Jeffersonville, IN)

ICELANDIC SHEEP

These sheep were brought to the island by Viking settlers, and few sheep have been imported to Iceland since settlement there ended about 900 years ago, leaving this breed one of the purest in the world today. Its first importation to North America in the twentieth century was in 1985, into Ontario, Canada.

The Icelandic is of the North European Short Tail type and related to the Finnsheep, the Romanov, and the Shetland, but is the biggest of that race, with good conformation for meat production. While raised in Iceland primarily for meat, outside of its native country the breed is best known for its wool, mostly marketed as Lopi yarn.

The fleece is dual-coated with an outer coat that can reach a length of 15 inches, and a shorter soft inner coat, with fleece in a wide range of colors with altogether seventeen main colors possible and at least twenty-four known markings.

This breed is best suited to a small farm, as their flocking instinct is poor. They are not docile, but alert and aggressive, showing great determination in going after their feed. The lambs, even though born small, go eagerly after mother's milk. The ewes are good mothers and protect their lambs well if threatened. On reasonable pasture, lambs reach good finishing weight of 100 pounds in three to four months.

Icelandic ewe. (Yeoman Farm, Parham, Ontario)

The meat is fine textured and has delicious flavor, while the wool is sought after by handspinners, and the skins make beautiful rugs.

The Icelandic is accepted by the Canadian Sheep Breeders Association and by Agriculture Canada as a pure breed and is registrable with the Canadian Livestock Records.

JACOB SHEEP

The Jacob sheep, once thought to be in danger of extinction, has now become so popular and relatively common in England that they have their own breed society, the Jacob Sheep Society, 242 Ringwood Road, St. Leonards, Ringwood, Hampshire BH24 2SB, England. While this society will allow the registration of animals with only two horns, most Jacob breeders feel that the special four-horn trait is highly desirable. The mottled fleece gives the Jacob sheep a unique and distinctive appearance, and has great appeal for handspinning wool. Tanned pelts bring premium prices.

KERRY HILL

This British breed originated in the Kerry Hills of Montgomeryshire, early in the 1800s. It is a hardy sheep and does best in hilly country, but is raised successfully even in lowlands and marsh areas.

The Kerry Hill has black and white markings on both face and legs, and a very black nose. The ewes are prolific and robust, and the rams noted for their olfactory powers, important to detect estrus in ewes who show only faint signs, as in out-of-season breeding. The rams are used in Wales for crossbreeding with the Welsh mountain ewes, producing a speckled face sheep that is very popular there. It is a good all-purpose breed for both meat and wool.

LEICESTER

The English Leicester has wool over the crown of the head, and resembles the Lincoln, except it is smaller, with a wool tuft on its forehead.

The Border Leicester has no wool on the head, less depth of body, a definitely Roman nose, and a stylish look. Their fairly erect ears seem far back from the wedge-shaped face. This is a long-wool breed, with coarse curled wool, 40s to 48s grade. This long lustrous wool is beautiful when dyed, and natural dark colored Border Leicester is much in demand among spinners.

LINCOLN

The Lincolns are from Lincolnshire in England, and are the largest of the sheep breeds, and slow maturing. Their long fleece is dense and strong and heavy, and they have forehead tufts. The breed is fairly hardy and prolific, but the lambs need protective penning for the first few days.

The Lincoln is not an active forager, and is best adapted to an abundance of pasture and supplements.

They do not stand cold rainy weather too well, as their fleece parts down the middle of the back, allowing the cold to hit their backbone, a sensitive area on sheep.

Jacob Ewe and newborn lamb. (Prairie Mary's Acres)

This is J.D. Paterson's champion Border Leicester ram at Ellesmere. (*New Zealand Farmer* magazine)

The author's favorite sheep, Mary, a charcoal-gray Lincoln.

Their fleece, however, is resistant to the deterioration shown in the wool of other breeds when parting along the back. It is sought by handspinners for the special long-wearing qualities and lustrous appearance. We like the wool for handspinning into almost indestructible sock yarn, and blended with other wools to make a strong weaving warp, with an attractive sheen.

MERINO

The Merino sheep, so famous for their fine wool, originated in Spain. They are descended from a strain of sheep developed during the reign of Claudius, from A.D. 14 to 37. These Tarentine sheep of Rome were crossed with the Laodicean sheep of Asia Minor by the Spanish. The Merino is one of the world's most popular sheep, and most wool breeds are originally part Merino.

Their fleeces are heavy in oil and, like the Rambouillet, lose much of their weight in washing. Since their lambs are small and slow-maturing, the main income is from the fleece and breeding stock for crossbreeding.

Not too long ago, the Merino was described as one of three types, A, B, and C. Type A had excessive skin folds, but this type has pretty well had the folds bred out, and Type A could be considered almost extinct. The folds and wrinkles of the skin were originally encouraged to give more skin area, and thus to have more area of wool follicles, and more wool production, but this wrinkling feature of the Merino is no longer considered desirable. The Type B Merino had fewer folds, and the Type C Merino, known as the *Delaine Merino*, *American Merino*, or just Delaine, had even fewer folds and wrinkles.

Merino rams. Ram on the left is a Superfine Merino ram from Australia and the others are Delaine Merinos. (Morehouse Farm, Red Hook, NY. Photograph by Francis J. Twomey.)

Most of the Merino sheep in the United States now are Delaine type, whose lambs grow faster than the old Type A and B. While the Merino ewes are not known as prolific milkers, they are very good mothers. Because this breed has the typical fine-wool trait of herding closely and can travel far for feed and water, Merino sheep are good on open range.

Australian Merino wool is classed in one of four categories: super-fine, fine, medium, and strong.

MONTADALE

The Montadale is an American breed, originating about 1932 in the St. Louis area, a cross of Cheviot rams and Columbia ewes.

The small head eliminates many lambing problems, and they are prolific lambers and good mothers.

They have heavy fleeces with little shrinkage, open faces, and clean legs. Wool grade is 3/8 blood.

The Montadale Breeders Association was organized in 1945, and by 1974 there were more than 68,000 registered.

The Montadale is easy to recognize, with a lot of style, a beautiful face, and those alert Cheviot-like ears.

OXFORD

The Oxfords originated in Oxford County, England, where they were bred from a primarily Cotswold and Hampshire foundation, which makes them a large heavy breed, with a good fleece weight and good length of wool (medium wool, 46s to 50s).

Champion Montdale ram.

The fleece is valuable for handspinning. They were successful in combining the hardiness, muscling, and wool quality of the Hampshire with the great size, growthiness, and wool quantity of the Cotswold and were first recognized as a true breed about 1862. They were imported into this country as early as 1846, and in expanding numbers into the 1900s.

The 1930s saw an overall trend toward a more compact type of sheep throughout the industry, and Oxford-sired lambs were criticized for being too large and grow-

Oxfords, owned by Mrs. Dan Korngiebel, Cuttingsville, VT.

thy, and not carrying enough fat as desired market weight. Oxford raisers were pressured to reduce the size of the breed to conform to the popular demand of the times. By the late 1950s, the outlook on fat in relation to lean edible meat had changed, and breeders looked again to larger and more efficient and productive sheep. As not all Oxford breeders had gone along with the reduction in body size, there was available stock to again "improve" the breed by regaining its large size and muscling. These are called *Modern Oxfords.*

It is most valuable as a sire breed. Rams now weigh up to 300 pounds, and in crossbreeding they contribute size and muscling to the resulting lambs. Being easily handled in small pastures, Oxfords are well suited to farm raising, and thrive with good feeding. The ewes are docile and heavy milkers. Since the breed has a head rather small for the body, the lambs are born easily.

Their faces and legs are usually light brown, but anything from light gray to dark brown is now acceptable, and a white spot on the end of the nose is quite common. With only a partly wooled face, there is no tendency to wool blindness.

Panamas, owned by Fred M. Laidlaw, Inc., Carey, ID.

PANAMA

The original Panama stock is the reverse of the parent breeds of the Columbia. Breeder James Laidlaw wanted to develop a sheep to replace the small Merinos that were very common in Idaho, and mated Rambouillet rams to Lincoln ewes. The aim was to get a more rugged sheep, with finer wool and better herding instinct than from the opposite cross of Lincoln rams on Rambouillet ewes. He felt that the ram had more influence on the offspring than the ewe, though this is still a matter of some controversy.

The first cross was made in 1912, starting with 50 rams and 1,600 Lincoln ewes. With this large number of rams and ewes, the breeders were able to avoid the inbreeding problems that sometimes arise in the attempt to form a new breed. The Rambouillet rams were only used for three years, and after that only Panama rams were used. By the end of five years, 1,000 Panamas were selected for the herd, and the

Perendale ram, owned by Norlaine Schultz.

remaining Lincoln ewes were sold off. After the first few years, the breeders started selecting for polled rams, and soon the breed was pure polled (hornless) as it is today.

The American Panama Registry was started in 1951, and all registered Panamas must be direct descendants of the original Laidlaw flock. They are good-sized sheep, hardy, good mothers, good milkers, with heavy fleece. A ewe fleece weighs 11 to 14 pounds, and is about 3/8 blood wool grade.

PERENDALE

The Perendale is a cross of Cheviot rams on Romney ewes, developed in New Zealand. They were first imported to the United States in 1977, and are growing in favor, being well suited to hilly areas.

Perendales have clean faces and legs and dense wool of four-to-five-inch staple which is unusually white, making it prized by spinners who dye their wool. It is easy to spin, and a good garment wool, a fine crossbred wool, 50s to 54s.

While this breed is easy to care for, and lambs unassisted, the animals inherit a bit of nervousness from their Cheviot ancestry and need gentle handling.

POLWARTH

The Polwarth breed has a history dating back to the 1880s, when it was bred to meet environmental conditions of Western Victoria, Australia, where the climate is too cold and wet for pure Merinos. It is a first cross Lincoln/Merino ewe mated with a Merino ram. Progeny from this are then mated. It was first known as the *Dennis comeback* and later *Polwarth* after the county in which it originated, in accordance with English practice.

The breeders' association was formed in 1919, and defines Polwarth as a dual-purpose sheep, approximately 3/4 Merino and 1/4 Lincoln, inbred to a fixed type with the emphasis on fleece. Fleece is about 58s to 60s count, not less than 4-inch length, dense and even, and carried well down on the belly. The wool-free face eliminates the need for facing (face-shearing is called *wigging* in Australia) and prevents wool blindness. Ewes will take rams at any time of the year, and have been successfully lambed twice a year, with high twinning. They are excellent mothers, lamb easily, and are good milkers.

The Polwarth has developed into an important breed in Australia, and has been exported to New Zealand, China, Nepal, Taiwan, Brazil, Uruguay, Argentina, the Falkland Islands, Peru, Pakistan, South Africa, Kenya, Korea, and recently to the United States.

POLYPAY

The Polypay is a new breed developed at the U.S. Sheep Experiment Station in Idaho, announced as a breed in the spring of 1976. It started with initial crosses of Targhee x Dorset and Rambouillet x Finnsheep breeds. These crosses were then crossed to form a four-breed cross. The lines were then closed with intensive within-line selection for lamb production when given the opportunity to breed twice a year.

It is a superior lamb production breed, with a quality carcass, and is giving outstanding performance in twice-a-year lambing, and also superior performance for conventional lambing. Only spring-born ewes which lamb at one year of age are kept for replacements.

Polypays. (U.S. Sheep Experiment Station, Dubois, ID)

The Rambouillet and Targhee breeding is included to retain hardiness and breeding season. Dorset breeding contributes to carcass quality, milking ability, and long breeding season. The Finnsheep breeding promotes early puberty, early postpartum fertility, and high lambing rate.

Careful and intensive selection has been used to increase further the ability of this new breed to produce at a high level, and the breeder certification program requires a careful mathematical production index to be kept on every ewe and ram.

The fleece of the Polypay averages 1/2 blood wool grade, with an average ewe fleece weighing about 8 pounds, and being very clean wool because of slatted-floor barn confinement during lambings and breedings. Wool weight would be higher if these ewes were not gestating and lactating twice each year.

The Polypay is another success from the same sheep experiment station that perfected the Columbia breed.

RAMBOUILLET

The Rambouillet is the French version of the Merino, developed from 386 Spanish Merinos imported by Louis XVI in 1786, for his estate at Rambouillet, and crossed with his own sheep there. They are very large and strong bodied, with very little wrinkling in the modern Rambouillet, except perhaps a little across the brisket. The fleece is less oily than the Merino, so has less shrinkage.

They are hardy, with a remarkable herding instinct. They graze during the day, and at night they will gather closely to sleep. They are good for open range, and can adapt to a wide degree of climate and feed conditions.

The ewes can be bred early, to lamb in November and December, and the lambs give good yield in boneless trimmed meat cuts.

They are a dual-purpose sheep, with a desirable carcass and good wool production. The rams have horns, and both sexes have white feet and open faces.

ROMANOV

The Romanov, like the Finn, is a northern "rat-tailed" breed. It first arrived in North America in 1980 when Agriculture Canada imported 14 ewes and 4 rams from France. After five years in quarantine, the Romanov stock was then released.

It appears that the Romanov's fertility, body size, growth, and carcass characteristics are similar to those of the Finn breed. The lambs are born black, with a silky hair coat over their wool. As they mature, the hair is shed and replaced by wool. This breed has been raised for its fur or pelts, in Russia.

One Romanov advantage is the early sexual development, maturing at six months of age. Ewe lambs first give birth when they are eleven or twelve months old. They also have the ability to breed out of season, and in Canada are producing lambs every eight months.

The herding instinct of the Rambouillet is shown here. (American Rambouillet Breeders Association)

Romanov breed. (Photo from Agriculture Canada Research Station, Lethbridge)

A Romeldale ewe and her twins. (Hank Saxton flock)

ROMELDALE

The Romeldale is a cross of Romney rams and Rambouillet ewes, producing 1/2 blood wool with very little shrinkage, making more pounds of clean wool than fleeces of fine wool. The straight Romeldale lamb is very marketable, and also one that can be saved for flock replacement ewes.

This breed is found mainly in California, and its popularity has not spread because of an inactive breed association and registry to publicize it, and also a general unavailability of good Romeldale rams.

Hank Sexton's flock in Willows, California dates back to the 1920s, and his flock improvement system to maximize ranch profits by increasing pounds of quality lamb and wool produced per ewe and per acre has been written up by the University of California Extension Service as a model for this type of program. The local Agricultural Extension agent says that Mr. Sexton has increased the ratings of his Romeldales at least 1/2 point in all classifications, over the breeds chart ratings at the end of this chapter.

The Romeldale is slightly smaller than the Columbia.

Handspinners are excited about the establishment of a breed registry for the newly developed CVM (California Variegated Mutant) Romeldale, as this will assure the pure breeding of the stock and an available source of the spinning fiber supply.

ROMNEYS

The Romney is an English breed, there called the Romney Marsh, after a low marshy reclaimed area where they are thought to have originated. They are said to have hooves less prone to the diseases of wet weather, and be somewhat resistant to liver flukes, another danger of damp pastures.

Romneys have a quiet temperament, can do well on a medium-good pasture, but are not suited to hilly country or hot, dry climates. They have little herding instinct, but are easily managed.

Romney ewe and lamb.

They have a finer fleece than the other long-wool breeds, but long and lustrous, almost as fine as some of the medium-wool sheep. Except for a tuft of wool on the forehead and short wool on the lower cheek, the rest of the head is clean. Their fleece does not have a tendency to part along the back, so they do well in rainy weather. It is an excellent handspinning fleece. Many are raised in Oregon.

They are a good quality meat animal, with a delicate taste. Recent studies have indicated that wool grade and flavor are actually related. Taste tests at the University of Idaho, using lambs raised at the U.S. Sheep Experiment Station in Dubois, found that as breeds go from fine to coarse wool, the amount of "muttony" flavor in the meat decreases as the wool grade gets coarser. This may explain why now the Romney has almost replaced the Merino among the flocks of New Zealand.

SCOTTISH BLACKFACE
(Black Face Highland)

This breed originated as a mountain sheep in Scotland, and it is a hardy quick-maturing meat animal. It has a lightweight fleece of long coarse wool. Both sexes have horns. In addition to an attractive and stylish fleece, its Roman nose and unusual black-and-white face markings set it apart in appearance. The mottled faces are preferred over the solid color black face in some parts of England, where the markings are said to indicate greater disease resistance.

SHETLAND

Canada imported a group of Shetland sheep in 1980, after a year's quarantine in the British Isles. The original sheep were in lifetime quarantine in Canada, but the progeny were eligible for release at five years of age. All animals were inspected regularly by Agriculture Canada, but no serious health problems were found. The first large Shetland flock was brought to Vermont in 1986 by Linda and Tut Doane (Roxbury, VT).

An ancient breed, the Shetland rams have two horns, while ewes are hornless, and all have short tails that do not require docking. Their wool is fine (64s/66s) but more durable than Merino and less likely to pill. A great range of natural color adds to its value, especially for handspinners. The many fleece colors include sparkling white into iridescent Sholmit and Shaela (shades of gray) to lustrous black and through Eesit (tan), Mooskit, and Mogit to Moorit, which is a deep, dark, rich chocolate brown. Shetland wool is used for the traditional wedding shawls that can be pulled through the bride's ring.

SHROPSHIRE

Shropshire is one of the "down" sheep, developed in southern England in the low hills called "downs."

It is a medium-small-sized sheep, with good meat lambs, but needs abundant feed. First known as a fixed breed in 1848, it was imported into the United States in 1855, and became well established as a popular breed. Its wool is 3/8 to 1/2 blood, with average fleece weight of about 10 pounds.

The distinctive Scottish Blackface. (*Sheep Breeder and Sheepman* magazine)

Shetland ram, owned by Linda and Tut Doane.

A modern type Southdown with New Zealand blood lines, owned by George A. Downsbrough, State College, PA.

Suffolk ewes. (E. William Hess)

The excessive face covering so encouraged in Shropshire breeding in the past has recently become unpopular because of the high incidence of wool blindness, and is now being bred out.

SOUTHDOWN

The Southdown was at one time the favorite meat breed, with a medium-small size, but good weight. It has a short broad head, dark legs, and face partly covered with wool.

It is one of the oldest English breeds, originating in the South Downs, a hilly portion of southeastern England, and is well adapted to grazing a hilly pasture.

A small, short and blocky meat lamb is not held in the same esteem that it was at one time, so the number of Southdowns appears to be dwindling, from a high of over 11,000 registered in 1958 and 1959 to only 3,558 in 1975. Actually, the 1975 figure is up about 200 from 1974, so the numbers may be leveling off now.

The Southdown fleece is a medium wool, 56s to 60s, and its short staple is suitable for handspinning into fine yarn.

SUFFOLK

The Suffolk is the most popular breed in this country, with nearly 47,000 new registrations in 1975. It is a handsome sheep whose black face, ears, and legs are free from wool.

The ewes are prolific and good milkers, with very little trouble at lambing. Their lambs grow rapidly, with more edible meat and less fat than many other breeds, plus a fine texture.

Although they placed fourth in the breeds chart, they rated highest in ewe size and ram size, highest in growth rate, feed efficiency, and muscling. They were topped only by the Finnsheep in ease of lambing and milking ability, in the breeds on that chart. They led all the other breeds in ewe size, ram size, and muscling. However, they lost points by being far down the scale in longevity and in wool production, having short wool with a fleece that is lightweight.

They are active grazers, able to rustle for feed on dry range, and they travel far to look for feed.

Originally an English breed, it was developed by the breeding of the dark-faced Southdown with the old Norfolk sheep, a black-face horned sheep that was hardy and prolific with meat of a superior texture but with poor conformation. The resulting breed combined the best of these parent breeds, with growing popularity in both purebred herds, and in usefulness for crossbreeding. Suffolk rams are used to cross on Rambouillet range ewes, to obtain the desirable Suffolk qualities in their lambs.

Records from 1920 show the registered Suffolk number at 805 in this country; in 1975 alone over 45,000 Suffolk were registered.

TARGHEE

The Targhee is a hardy American breed, developed by mating outstanding Rambouillet rams to ewes of Corriedale x Lincoln Rambouillet stock, and ewes of only Lincoln-Rambouillet, and following that by interbreeding the resulting lambs. This work was done since 1926 by the United States Sheep Experiment Station in Dubois, Idaho, to meet the demand for a breed of sheep that was thick in natural muscling, prolific, producing high-quality apparel-type wool, and adapted to both farm and range conditions. It gets its name from the Targhee National Forest on which the Experiment Station flock grazes in the summer.

It is a large-sized, dual-purpose sheep with a good meat type and heavy fleece (11 to 16 pounds) of good wool, about 1/2 blood, 3 inches length or more. It has a clean face and no skin folds, with ewes weighing from 125 to 200 pounds, and rams from 200 to 300 pounds.

Experimental work at the University of Idaho has shown the Targhee to have an inherited resistance to internal parasites and to hoof troubles. The breed can also claim a very long productive life.

It is noted for an ease of lambing, and high percentage of twins or triplets.

On the National Livestock Producer's breeds chart, shown at the end of this chapter, it took second place.

TEXEL

The Texel of Holland has received considerable attention from breeders, but because of a long-incubation-period pulmonary disease prevalent in this breed, its importation has been difficult. However, in 1985 the USDA's Agricultural Research Service imported four Texel rams and twenty ewes from Finland, along with four Texel rams from Denmark. They were more easily imported from Finland and Denmark because these countries are free of foot-and-mouth disease. These sheep will remain in quarantine for at least five years, with USDA scientists conducting research

Yearling Targhee ewes from the U.S. Sheep Experiment Station, Dubois, ID.

in the basic biology of lean meat production, and evaluating Texel as a breed to sire lambs from crossbred ewes.

The Texel has many genetic advantages, such as high fertility, large size, rapid growth rate, excellent lean carcass, and a high wool production of fine crossbreed fleece.

This breed resulted from many crosses of the native "polder sheep" (grazers on polderland reclaimed from the sea) with British breeds such as Border Leicester and Lincoln to get a better meat animal, which was then crossed back to rams of the original breed.

Texels adapt easily to new environments, with many selling from Holland into South America. While prolific, it lambs only once a year, and is not ideal for open range use as it does not have a good herding instinct.

TUNIS

The Tunis is one of the oldest of the distinct sheep breeds, dating back over 3,000 years. The first importation into this country was in 1799, with sheep from the flock of the Bey of Tunis, in Tunis, Africa, being brought to Pennsylvania. From there they spread mainly to Virginia, Georgia, and the Carolinas.

A Tunis ram was used by George Washington to rebuild his flock, which had declined in numbers and vigor while he was serving as President. The Tunis could have been a major breed in this country, had not most of the southern flocks been destroyed during the Civil War.

They are medium size, hardy, docile, and very good mothers. The ewes are known for breeding out of season, and with proper management they can be bred

Tunis yearling ewes raised by Jim Lillie, Fitchburg, MA.

almost any month of the year. An unusual color of reddish tan hair covers their legs and faces and their long broad pendulous ears. They have a medium-heavy fleece of 1/4 to 3/8 blood, very good for handspinning. The lambs are a reddish color when born and gradually lighten to white, although retaining the distinctive red-tan on their legs and face.

The Tunis does well in a warm climate, and the rams remain active in very hot summer weather. Although they are a superior breed for a hot climate, they are raised successfully almost anywhere, and their concentration here is mainly in the north-eastern states.

With the current interest in out-of-season breeding, prolificacy, and milking ability, the Tunis is again increasing in numbers, as well as attracting attention in cross-breeding research projects.

WILTSHIRE HORN

The Wiltshire Horn sheep is an ancient British hair-sheep breed, once known as the "Western."

In England the rams are used for crossbreeding with smaller breeds of ewes to obtain fat lambs. The lambs inherit the increased fattening quality of the Wiltshire,

Breed	Ewe Size	Ram Size	Growth Rate	Feed Efficiency	Muscling
Black-Faced Highlands	3.0	3.0	3.0	3.0	3.0
Cheviot	2.0	1.8	2.3	2.7	2.0
Columbia	4.8	4.8	3.3	3.5	3.5
Corriedale	3.3	3.3	3.0	3.0	2.8
Cotswold	5.0	5.0	3.3	2.7	3.0
Debouillet	2.5	2.5	2.5	2.0	2.0
Dorset	3.0	3.3	3.3	3.3	3.3
Finnish Landrace	2.2	2.2	3.3	3.3	2.6
Hampshire	4.6	4.6	4.6	4.3	4.6
Karakul	2.5	2.5	2.5	2.5	2.0
Large Border Leicester	4.3	4.3	4.0	3.7	3.7
Lincoln	4.5	4.5	3.3	2.7	3.0
Merino	2.3	2.3	1.7	2.0	1.7
Montadale	3.0	3.0	2.7	3.0	2.7
Oxford	4.0	4.0	4.0	3.3	3.7
Panama	4.3	4.3	3.7	3.0	3.7
Rambouillet	4.0	4.0	3.5	3.3	3.6
Romeldale	3.0	3.0	3.0	3.0	3.0
Romney	3.0	3.0	3.0	3.0	3.0
Ryeland	3.0	3.0	3.0	3.0	3.0
Shropshire	3.5	3.8	3.0	3.0	3.0
Suffolk	5.0	5.0	4.8	4.8	5.0
Southdown	1.2	1.2	1.2	2.5	3.0
Targhee	4.3	4.3	3.8	4.0	3.5
Tunis	3.0	3.0	3.0	3.0	3.0

and can be brought to market on pasture alone. The lambs also inherit the narrow head, which helps in lambing.

This breed has been dwindling rapidly, and also is being diluted by crossbreeding. Until recently they were included with the sheep breeds to be preserved by the Rare Breeds Survival Trust of the Royal Agricultural Society in England, dedicated to the protection of breeds that are in danger of extinction.

Pictured here are the first Wiltshire Horn sheep to arrive in Canada, at the ranch of R.A.K. Turner, in Nova Scotia (1972).

WHICH BREED IS BEST?

This is the charted result of a survey conducted by Frank Lessiter, editor of the National Livestock Producer, with breeds rated by seven prominent sheepmen, as shown in the October 1975 issue, reprinted by permission.

---

* Denotes tie.

The breeds were compared as to their potential value to a commercial sheepman—with a flock of 250 grade ewes—who was willing to crossbreed. The breeds were scored by seven sheep experts as follows: 5.0—Excellent; 4.0—Good; 3.0—Average; 2.0—Unsatisfactory; and 1.0—Poor.

Wool Production	Wool Grade	Out of Season Breeding	Ease of Lambing	Milking Ability	Longevity	Hardiness	Overall Breed Ranking
3.0	2.0	2.5	3.0	2.5	3.5	2.5	22nd
1.7	3.0	2.0	4.0	2.0	3.0	3.8	23rd
4.2	4.0	2.4	3.3	3.5	3.0	3.4	3rd
4.3	4.0	2.7	3.7	3.7	4.9	3.7	7th
3.0	2.0	2.3	2.3	2.3	2.7	1.7	19th
5.0	5.0	3.5	4.0	3.0	4.0	4.0	9th
2.0	3.0	4.3	4.0	4.0	3.0	3.0	10th
1.6	2.0	4.0	5.0	4.4	2.7	3.0	15th
1.8	2.7	3.0	2.7	4.0	2.3	3.0	5th
2.0	2.0	2.5	2.5	2.5	3.0	2.0	24th
3.8	3.3	3.0	3.0	3.0	3.0	3.0	6th
2.8	2.0	2.3	2.3	2.3	2.7	2.3	21st
5.0	4.7	3.0	3.0	2.3	3.7	4.3	16th*
3.0	3.0	2.7	3.0	3.0	3.0	3.0	20th
2.7	2.7	2.3	3.0	2.7	2.7	3.0	11th
3.7	3.0	3.0	3.0	3.0	3.0	3.0	8th
5.0	5.0	4.2	3.8	3.4	4.8	4.8	1st
4.0	3.5	3.0	3.0	3.0	3.0	3.0	12th
3.5	3.0	3.0	3.0	3.0	3.0	3.0	14th
3.0	3.0	3.0	3.0	3.0	3.0	3.0	16th*
3.0	3.0	2.3	3.3	3.0	3.3	3.0	13th
1.0	2.0	3.3	4.0	4.4	1.6	2.5	4th
1.8	2.3	2.0	2.7	1.7	3.3	3.0	25th
4.5 ·	4.7	3.0	4.3	3.8	4.0	4.3	2nd
2.5	2.5	3.0	3.5	3.5	3.0	3.0	16th*

EWE COMPARISON

Breed	Heavy Weight When Fat (Lambs)	Grazing Ability	Heat Toler- ance	Prolif- icacy	Mother- ing Ability	Temper- ament
Black-Faced Highland	3.8	5.0	3.0	3.0	3.5	3.0
Cheviot	1.5	5.0	3.5	1.0	5.0	1.0
Columbia	5.0	4.2	4.5	3.5	4.0	3.8
Corriedale	3.5	4.0	4.5	3.0	3.2	3.6
Cotswold	5.0	1.5	2.5	3.5	2.5	3.0
Debouillet	3.0	4.5	5.0	3.0	3.0	3.0
Dorset	3.0	3.2	4.0	3.5	4.5	4.0
Finnsheep	3.5	3.0	3.0	5.0	4.5	2.5
Hampshire	4.5	3.8	4.0	3.2	4.5	4.0
Karakul	3.0	4.0	4.5	2.0	2.5	2.0
Border Leicester	4.8	2.8	3.0	3.5	2.5	3.0
Lincoln	5.0	2.0	3.0	3.5	3.0	4.0
Merino	1.5	5.0	5.0	1.0	2.0	1.0
Montadale	3.0	4.5	3.5	3.5	4.0	1.0
Oxford	4.0	3.0	4.0	4.0	4.0	4.0
Panama	5.0	4.2	4.5	3.2	3.5	3.5
Rambouillet	4.2	4.5	5.0	3.0	3.2	3.0
Romeldale	3.0	3.5	4.0	2.5	3.8	3.5
Romney	3.5	4.0	3.5	3.0	4.0	4.8
Ryeland	3.0	3.5	3.0	2.5	3.0	4.0
Shropshire	3.0	3.5	3.5	3.5	4.5	4.0
Suffolk	5.0	4.0	4.0	3.5	4.5	2.8
Southdown	1.5	1.0	1.0	2.5	4.5	5.0
Targhee	4.5	4.5	4.8	3.2	3.5	3.5
Tunis	3.0	3.5	5.0	4.0	4.0	3.0

These additional traits were rated by Robert M. Jordan, professor of Animal Science, University of Minnesota, and some of these were printed in *Shepherd* magazine in March, 1976.

Some traits are of importance in one sex and not visibly important in the other, except as the trait is passed on to the offspring. Examples are prolificacy and mothering ability of the female, and heat tolerance, which is primarily of interest in the male by way of sterility in very hot weather.

The ratings of many of these traits are not permanently fixed; they can still be improved by breeding and selection. This is most noticeable in prolificacy, for if you retain only twin ewes for breeding purposes, and use rams that are also twins, you can increase your lambing percentages above what might be expected of a less prolific breed.

Management and nutrition also play a large role in the exploiting of the potentially profitable traits in *any* of the breeds.

CHAPTER 3

Fences and Pastures

MORE THAN 90 MILLION ACRES in this country cannot be used efficiently for anything other than growing grass, and some parts of all farms are probably less suitable for crops than for pasture.

Sheep are more efficient than cattle in converting grass to meat. They have their lambs in the spring, so the lambs grow to market age on the abundance of summer grass and can be sold about the time the pasture gives out in late summer or early fall. This means that you do not need to carry the meat animals through the winter on hay and grain, as you would beef animals.

PASTURES

STOCKING RATE

How many sheep can be kept per acre? A number of factors are involved in deciding this question. These include: type of soil (rock, sand, clay, etc.); plant species (grass, weeds, clover, etc.); rainfall or irrigation; climate; fertility of soil; lay of the land (hill, slope, marsh); whether lambs, ewes with lambs, or dry ewes; and whether pasture can be rotated.

The fewer breeding ewes grazing during the winter, the less supplemental feed they will need at that time. Sheep do not do as well when pasture is overstocked, and the older ewes suffer the most. Their poor teeth make it harder for them to chomp on overgrazed pasture, and with short grass they obtain less feed per bite. Even the teeth of younger ewes suffer from having to crop the short grass, for with it they are getting a certain amount of dirt and sand.

Some farms estimate four sheep to an acre of good pasture, with hay and some grain in the winter, and one or two sheep to an acre of poor pasture, again with supplemental food in winter. So you should take a good look at the condition of your acreage. Better to keep too few the first year, and see how the pasturage holds up.

ROTATION

Pasture rotation serves to maximize the number of sheep that a given area of forage will support, and is also an aid in the control of parasites. When you are confining them at night, sheep tend to graze out from the barn during the day, then back again. Thus an area close to the barn becomes as bare as a parking lot, whereas the farthest reaches of the pasture are undergrazed and become rank and overgrown.

The technique of pasture rotation is simply to divide the pasture into smaller pastures, with electric fence or other light fencing, so that each smaller pasture is "over-stocked" and can be grazed completely in ten to fourteen days. The sheep will graze this pasture more evenly and have less tendency to pick and choose as they would in larger pastures. When the area has been grazed, the sheep are moved to a fresh pasture. If you have three or four small fields, the grass will receive several weeks of rest between grazings, which will also allow time for many worm larvae to die of exposure and old age. Scottish shepherds say, "Never let the church bell strike thrice on the same pasture."

Some plants, such as alfalfa, cannot stand continuous grazing, but can stand hard use over shorter periods of time. Alternating pastures gives the plant a chance to recover, and gets more actual growth out of the same amount of space. Most grasses grow better when given a periodic rest from grazing.

On large farms, it has been determined that one hundred sheep will do better on twenty-five acres that is divided than on thirty acres with no division. This holds true on smaller farms, too.

Sheep by nature would prefer not to feed continually in the same place. They like a fresh pasture that has not been recently walked on. If with ample pasture they seem choosy and walk around with their noses to the grass but not eating as much or chewing their cuds as much, it is time for a pasture change. They can even be moved to one where the grass is not much higher. It will have "fresh" grass, and they will eat better.

Perennial grass stores food in its roots after it has made the season's main growth. The grass uses these reserves to survive while dormant, to make the first spring growth, and to start new growth after its leaves are closely grazed. Its ability to recover quickly after grazing makes grass valuable for forage production, but can deceive us into thinking that leaves can be repeatedly removed without injury. If they are, the plant keeps drawing on food stored in the roots to grow new leaves until the supply is exhausted and the grass dies.

Many grasses will not reach their maximum vigor and growth when more than half of their leaf surface is removed. This weakened grass does not make efficient use of soil moisture and nutrients, and does not provide the maximum livestock feed.

During the droughts of the 1930s and 1950s, wind erosion occurred largely on land that had little or no plant cover because of cultivation or too close grazing. If you rotate your pastures and prevent excessive over-grazing, grass pasture will not only feed the sheep, but will also protect the soil from erosion by wind or water.

Because of the added expense and labor, rotation fences are often neglected. However, cross fences need not be as heavy or as high as the outer fence, nor need they be dog-proof.

When rotating pastures, don't let the grass get too tall before you turn the sheep in, for they will trample more of the grass and will not eat it as well as they would shorter grass. From 4 to 6 inches height is ideal.

SHEEP AND GOAT PASTURE

If the pasture was cleared once, but has been without livestock for a while and is partly grown up with blackberries, Scotch broom, small saplings and brush, you can run goats with the sheep to clear the brush. While sheep like to eat from 4 inches to 8 inches from the ground, goats like to eat from about 10 inches to as high as they can reach, and are great brush and bramble clearers. Sheep are grazers, goats are browsers.

ORCHARDS

An orchard is one of the favorite sheep pastures on a small farm. The sheep can make use of the shade in the summer, and if a little care is exercised to prevent them from getting too many windfall apples and other fruit, they can make good nutritional use of fallen fruit. They should not be turned suddenly into an orchard with a lot of fruit on the ground. However, if they are there all the time, they are accustomed to the fruit in their diet. The extra fruit should be picked up and taken away, so they don't get too much. This fruit will keep long enough to be added to their diet after the fallen fruit in the orchard is gone. We put away windfall apples and dole them out, a few at a time, far into the winter. Try to save some apples until lambing time, so they can be given as treats to the ewes when they are in the lambing pens.

You will find an occasional sheep with goat-like habits, standing on its hind legs and nibbling the branches and leaves on the fruit trees. Usually sheep don't do that, but they will chew on the bark of the trunks, and can do a lot of damage if you do not protect the trees. The trunks of larger trees can be wrapped with several layers of chicken wire, or once around with rabbit wire. Old burlap feed bags can be used, and fastened with wire or baling cord. A temporary solution is to make "manure tea" from sheep droppings, and paint that on the trunks. This would need to be repeated after every rain.

Any small or newly planted trees will need the protection of a rigid fence. A board fence around small trees provides secure protection for the tender tree, and is easy for people to climb over to prune or pick fruit. Avoid having the fence so close to the tree that sheep will brace their feet on it to reach the lower branches.

If you buy an old farm, the fences and barn buildings will probably need repairs. The buildings can be repaired after you get the sheep, but the fences should be done before. Sheep quickly learn to jump sagging fences. One sheep loose in a neighborhood can be quite a problem, and a sheep in your garden is a disaster.

If you wait until they have the jumping habit, they may still do it after the fence is repaired. One jumper can set a bad example and should be sold, or slowed down by temporary "clogging" until retrained. Fasten a piece of wood to one front ankle with a strap. It will get in the way just enough to prevent the sheep from jumping.

POSTS

The life of any fence depends a lot on how hard the sheep worry it, and how long the posts, especially the end posts, hold up. They should be massive and solid, for if they start rotting and the sheep rub on them, or put their heads through the fence and strain to reach greener grass on the other side—there goes your fence.

Painting posts can make a fence more attractive, but not more durable. Posts tapered at the top to drain off rain and snow sound reasonable, but tests show no improvement in the life of the posts. The only reliable way to prolong the posts' usefulness is with wood preservatives.

ESTIMATED LIFE OF UNTREATED WOOD POSTS
(Round, 5 to 6 inches in size)

Over 15 Years	7 to 15 Years	3 to 7 Years	
Black Locust	Cedar	Ash	Honey Locust
Osage-orange	Red Cedar	Aspen	Maple
	Red Mulberry	Balsam Fir	Pine
	Redwood	Beech	Red Oak
	Sassafras	Box Elder	Spruce
	White Oak	Butternut	Sycamore
	Douglas Fir	Tamarack	
	Hemlock	Willow	
	Hickory	Yellow	
		Poplar	

Split posts, which have more "heartwood," will last longer than the time listed, and larger sized posts also last longer. In addition, the life of the post is doubled or even tripled when it is properly treated.

If untreated posts are set in concrete, or thinly coated with concrete, water can get in cracks between the post and the concrete, and moisture will be held in, so the posts will rot faster. Some farmers pile stones around untreated posts to keep back the weeds and promote air movement around the posts, but the stones hold moisture at the ground line, and encourage decay there.

> Do not use paint containing lead on sheep equipment or on parts of buildings accessible to sheep. Poisoning may result when animals constantly lick or chew objects covered with paint containing lead.

PRESERVATIVES

Coal-tar creosote is the best preservative. Since heat and special equipment are required, this is a difficult process for the home owner. There are many other ways of treating wooden posts to extend their usefulness, but not all are practical for the small farm, and safe for animals.

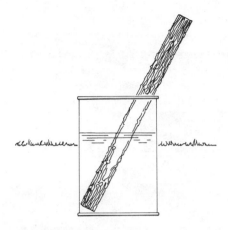

Oil drum used for fence post treatment.

END SOAKING

If you need a fence right now and don't have time to cut your posts ahead and season them before treatment, end soaking is the method to use. Cut round posts and leave the bark on the posts. Then use a 15 percent to 20 percent solution of zinc chloride, or chromatized (chromated) zinc chloride in water, (which is nontoxic to humans or sheep). Allow about 5 pounds (or about a half-gallon) of the solution for each cubic foot of post to be treated. The chromated zinc chloride is sold in a granular form that is easy to use, and is less subject to leaching from the posts than the plain zinc chloride. Often it is difficult to find a source for this chemical. If you can't find it, you usually can buy its two ingredients from a chemical company, and mix them. Use 80 percent zinc chloride and 20 percent sodium bichromate. For the solution mix a 20 percent chemical, 80 percent water solution (each by weight). Thus 100 gallons of water weigh about 830 pounds and would require 166 pounds of the zinc mixture.

Stand the posts bottom down in a tub or drum of this solution until they absorb about three-fourths of it, which takes from three to ten days. Then stand them on their tops, and let them absorb the rest. They should be seasoned for about a month before using, to allow the treated wood to dry.

Either green or seasoned posts may be soaked, covered completely, and steeped, in a 5 percent solution of this same zinc chloride, for from one to two weeks.

The best absorption and penetration are obtained by first seasoning the posts. This lets the sap dry out to make room for the preservative. Peeled posts should be open piled, so that the air can circulate around each one, and the bottom of the pile should be at least a foot above the ground. The best place for piling would be an exposed area on well-drained ground.

While posts cut in the spring will peel more easily, posts cut in the fall will have a chance to dry more slowly, which prevents some cracking and checking. This is more important with oak posts than with wood from cone-bearing trees.

The seasoning of posts adds little to their life unless they are also treated with preservatives.

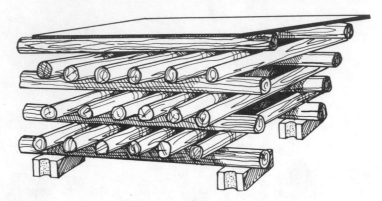

Proper storage of fence posts, for drying. Note "roof" on top of pile, to keep off rain.

COLD SOAKING OF SEASONED POSTS

The method of soaking posts in a solution of pentachlorophenol, called "penta" for short, was used at one time. Since penta is highly toxic and can be absorbed through the skin and by way of the lungs, its use is now completely discontinued—wisely so.

CHARRING

There is an old-fashioned method of preserving posts that costs nothing, uses no preservatives, and is quite effective. You just char the part of the post that goes into the ground, to at least 6 inches above ground, by placing that portion of the post in a fire—a bonfire or large wood stove. The object is to get about 1/4-inch minimum charring on the outside of the post, which seems to reduce the incidence of rot and extends the life of the post.

SALT TREATING

Most lumber yards sell "salt-treated posts," which have been treated with chromated copper arsenate, an effective preservative that is not palatable to livestock.

STEEL POSTS

The use of steel posts can save a lot of labor in digging and tamping dirt around posts. It is even more of a time- and labor-saver when you are fencing through a wooded area where there are roots and stumps. Driving a slender metal post is easier than digging a hole.

Steel posts are driven into the ground with the use of a 16-pound "sleeve" that fits over the top of the post. While it is suggested you pound it with a sledge or by

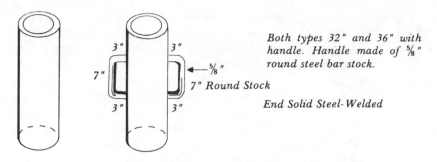

Both types 32" and 36" with handle. Handle made of ⅝" round steel bar stock.

3"   3"

7"   ⬅ ⅝"

7" Round Stock

3"   3"

End Solid Steel-Welded

Two styles of "sleeves" for driving steel posts.

hand, the latter— raising up and pounding down over the post—is the simple way. There are also heavier spring-loaded post driver sleeves, with handles on the sides.

Galvanized steel posts have the longest life, followed by posts brush-painted with metallic zinc. Posts that are dip-painted with lead and oil paint can start to rust within five to seven years.

Some posts are sold with five clips included in the price of each post. If not, order the clips at the same time as the posts, so you get the right style.

Old iron bed-rails where *springs* were fastened, often found in a salvage yard, make good posts. Some already have holes in them where the springs were fastened. If not, you may want to drill some holes to fasten the fence.

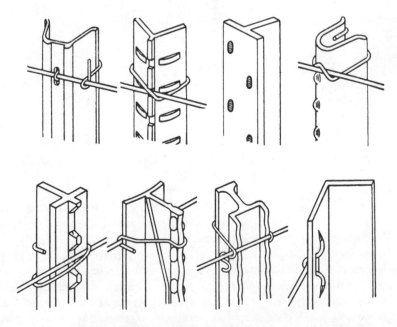

Steel fence posts and methods of fastening wire to them.

Discarded pipe from machine or repair shops is sometimes much cheaper than steel posts. It should be at least 1 3/4 inches in diameter, and heavier than that for corner posts. End posts of 6-inch or 8-inch pipe can be filled with concrete.

Concrete can be poured around the steel post, with the "form" dug in the ground. The hole-form should extend 3 1/2 feet into the ground, about 18 inches square at the top, and 20 inches square at the bottom. Posts should be maintained about 1 inch out of plumb, away from the direction the fence will be stretched, while pouring the cement. The braces for the steel corner posts should be attached to the corner posts before positioning the ends that will be sunk in concrete. Determine location of concrete pier, and dig form about 20 inches square, and 18 inches deep. The brace will enter the concrete about 6 inches below the ground, and extend at least 6 inches into the concrete.

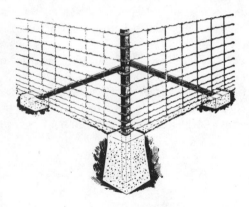

Steel corner and brace posts set in concrete.

DEPTH OF SETTING

Line posts are usually set 2 1/2 feet in the ground. End, corner, and brace posts are set 3 1/2 feet deep, and gateposts are set 4 feet deep.

Hand augers or posthole diggers are easier to use than shovels, and there's less dirt to put back into the hole. In heavy soil or clay, oil the posthole digger so the dirt does not stick to it. Keep a bucket of waste crankcase oil where you are digging, and dip the digger into it frequently. Posts should have the dirt tamped tightly around them, and should be aligned with the rest of the fence posts while being tamped. The longest-lasting posts are purchased ones, pressure-treated with coal-tar creosote, with an average life of at least thirty years, but initial costs are high, and they are not available everywhere. A fence is no stronger than its end posts and braces. The brace wire has its ends spliced together, and is tightened by twisting it with a strong stick or rod. Leave the stick in place so you can adjust it as necessary.

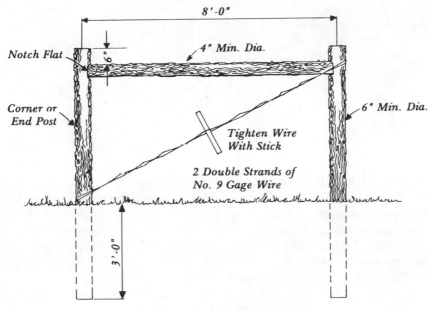

Corner or end post assembly.

FENCING

WOVEN WIRE FENCING

Woven wire fencing is the most practical fence for sheep, at least for the bottom half of the fence. Wire "stock fence" comes in different weights and styles. The five most common style numbers are 1155, 1047, 939, 832, and 726. The last two numbers denote the height of the wire in inches, and the first one or two numbers indicate the number of horizontal (line) wires in the fencing. There are three other styles, not seen as often: 949, 845, and 635. All the heights come in a choice of stays that are either 12 inches or 6 inches apart. While 6-inch stays will stop more dogs, the sheep don't get their heads stuck in the 12-inch stays. A small dog could possibly go through 12-inch stays, but a dog that small could probably get through lots of other places.

Woven wire is usually sold in 330-foot rolls, in a choice of four weights, although not all may be available locally. The weight depends on the gauge of the wires, and the lower the number given for the gauge, the heavier the wire: 9, 11, 12 1/2, and 14 1/2 are standard. The top and bottom line wires are heavier than the stay and intermediate line wires.

FENCE WEIGHTS

Weight	Gauge of Top and Bottom Line Wires	Gauge of Filler and Stay Wires
Light	11	14 1/2
Medium	10	12 1/2
Heavy	9	11
Extra heavy	9	9

Woven wire is coated with aluminum or zinc (galvanized). The zinc coating can be of several different thicknesses, indicated by government class numbers: 1, 2, or 3, class 3 being the heaviest coating. The lightweight fencing is suitable only for cross fences, will not last as long as a heavier wire, and is easier for the sheep to wear down. The heavy and extra-heavy wires are hard to work with, especially if they are in 6-inch stays, or very high. Most people settle for the medium weight, so there is more demand for it, and it is easier to find.

A height of 26 inches or 32 inches can be handled with relative ease, and you build up the height with barbed wires. Place the first wire 1 inch above the top of the woven wire, for if the woven wire sags even a little, that is where the sheep will get their heads through. Also, string barbed wire about 1 inch below the bottom of the woven wire, to discourage dogs. When you stretch woven wire over uneven ground, there are places where it doesn't touch the ground. You can make up this difference with barbed wire along the bottom, parallel to the contour of the ground. A third wire half-way between the top and bottom of the woven wire will help reinforce the fence by discouraging the sheep from rubbing on it, or putting their heads through.

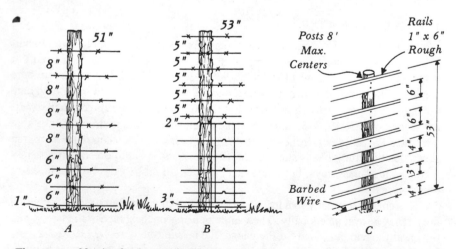

Three types of fencing for sheep. A is all barbed wire, B is a combination of fencing and barbed wire, and C is a board fence with barbed wire at the base.

Putting up fencing can be dangerous, but you can minimize the risk of injury by wearing heavy leather gloves, high boots, and tough clothing.

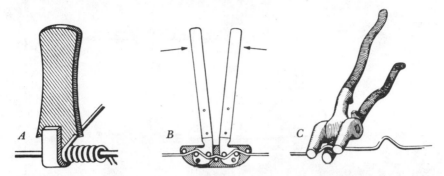

Useful fencing tools include (A) splicing tool, (B) double-crimp tool, and (C) single-crimp tool.

STRETCHING WOVEN WIRE

Woven wire can be stretched only from one anchored post to another, but an anchored post can be a gate, corner, end, or just a braced-line post—any but the unanchored line posts. For the sake of simplicity, we will assume that the fence is being stretched from one wooden end post to another.

Set the roll of fencing on end, about one rod (16 1/2 feet) from the end post. To prevent staples from being pulled by the sheep pressing against the fence, put the wire on the livestock side of the posts. Where appearance is important, you may want to put it on the outside.

Unroll enough wire to make a wrap around the post, and some extra. Remove two or three stay (vertical) wires, depending on the circumference of the post, and position the next stay wire along the edge of the post. Starting with the center line wire, wrap each horizontal wire around the end post and wind it back five times around the line wire.

Unroll the fencing, keeping the bottom wire close to each post. Before stretching the wire, tie it loosely to the posts with baling twine, or prop it up with temporary stakes. Place the clamp so that when the fence is stretched the clamp will pass beyond the other end post, enabling you to staple the fencing to the post.

You will only need a "single" stretcher on 26-inch or 32-inch woven wire. Some stretchers must be anchored to a dummy post or a tree.

To be avoided are the rope-and-pulley types that don't give enough control to avoid accidents, and require a great deal of pulling and tugging to get a fence tight.

Before you start stretching the woven wire, note the shape of the tension curves (the little crimps in the line wires). The fence is stretched properly when these tension curves are about half straightened out. This provides for expansion and contraction with changes of temperature, and also leaves tension in the fence, which makes for a tighter fence.

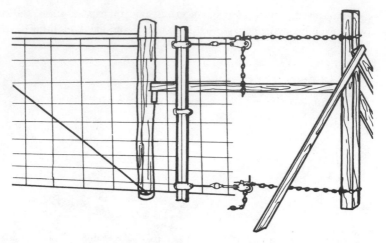

Fence stretcher in position. Note braced "dummy" post at right.

As a safety precaution when stretching, stand on the opposite side of the post from the stretcher so if something gives, you'll be out of the way.

Fasten the fence to the posts on the ridges and in the low places first, fastening the line wires one at a time, starting at the top or bottom, whichever is the tighter.

To counteract the tendency of the tightly stretched fencing in a low place pulling up the post, you can bury a "dead man," an old log, under your fence line at

A neat job of splicing woven wire. (USDA photo)

that place. Secure the bottom of the woven wire to the buried log, and it will do the work of a couple of earth-anchors, at no cost. A large boulder can be used in place of the log.

When fastening the woven wire to the end post, secure each line wire with two staples angled in opposite directions to prevent slippage. Measure the circumference of this post. Cut the fencing, allowing that many inches plus at least 6 inches more, for five wraps around the line wires. Remove a couple of stay wires, wrap the fence around the end post and secure it by wrapping the ends of the line wires back around the corresponding line wires of the fence. If the wire continues around a corner post, there is no need to cut it. Just fasten it and continue around.

Woven wire should be stretched and attached in sections running from one anchor post to the next.

BARBED WIRE

Barbed wire is usually available in both 12 1/2 and 14 gauge, with the lighter (14) used mainly for temporary fences. There are also two-point barbs and four-point barbs, but the number of barbs is more important if you are running cattle, where the complete fence would be made of barbed wire. It also has numbers to denote the thickness of the zinc or aluminum coating.

The handy "Barb Dollie," in position for pulling off barbed wire. Handle position can be reversed for easier moving of wire spool. This kind of equipment can be homemade.

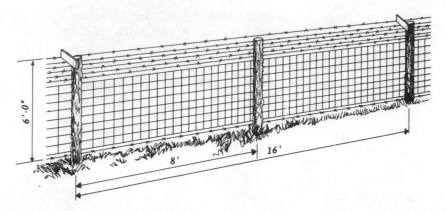

This combination woven fence and barbed wire is built to protect sheep from dogs and coyotes.

These instructions apply specifically to stretching barbed wire above woven wire. In such a fence, barbed wire is attached after woven wire is in place.

Before unrolling barbed wire, attach one end to the gatepost or end post. Two people can unroll a reel of wire by placing a rod through the center of the reel and letting it unroll as they walk down the fence line. (Nasco Catalog has a dandy Barb Dollie that makes it a one-person job.) Two or more reels can be unrolled by placing them on an end-gate rod in the back of a truck or wagon.

The rope-and-pulley stretcher, which I wouldn't recommend for woven wire, can be used to stretch barbed wire. But you can stretch it tight enough with a crowbar or a claw hammer so that any slack will be taken up when you staple it to the posts.

If you buy a farm with an existing barbed wire fence for cattle, the practical thing to do would be to add additional wires to make it more dog-tight and usable for sheep.

WOOD FENCES

Board, slab, or plank fences are good rigid fences for corrals, and attractive fences around a farmhouse. For corrals, the boards should be on the inside of the fence, so sheep will not loosen them. Posts would need to be closer together than for pasture fences, and the planks well spiked or bolted to them.

DOG-TIGHT FENCES

While sheep on open range and in less settled areas still have terrible problems with coyotes, in the populated places and small farm communities, the domestic dog does the damage. The U.S. dog population is estimated at over 22 million, which is about three dogs for every two sheep.

A dog need not be wild or vicious or even brave to chase a sheep. It is only following its natural impulse to chase whatever runs. Unfortunately, sheep startle and run at the slightest disturbance. The offending dog is not as much at fault as is the

Coyote snare being placed under fence. (*West Texas Livestock Weekly*)

owner who does not keep the animal at home.

A dog too small to kill or maim sheep can still cause death by heart failure in old ewes, and abortion in pregnant ewes. Broken legs, too, often result when sheep are chased.

For a dog-tight corral fence, the 12-inch stays are not suitable, and ones even smaller than the 6-inch would be best. We have used the "non-climbable horse fence" wire, and found that it is very satisfactory for a small area, but so rigid as to be hard to stretch unless you have a straight pull. Whatever you use should be extended up to at least 5 1/2 or 6 feet by barbed wire, to be a foolproof night corral. Don't neglect one strand of barbed wire at ground level, on the outside of the posts, to discourage digging. A corral fence of this type should provide a safe place to leave sheep at night, when most of the damage occurs.

Adding electric strands to an existing woven wire fence can also make a good dog-tight fence.

In addition to good fences, dogs can be discouraged by:

- The best solution could be a sheep guardian dog. A well-raised and trained dog can protect sheep against all predators (see Chapter 20).
- Night corrals, with high dog-tight fences, can protect during night and early morning hours, when most attacks occur.

- Bells on some of your sheep. You can hear the bells if the sheep are being chased. High-frequency bells have also been tried with some success, as dogs find the sound unpleasant.
- Having sheep in an open field, in sight of your house. This helps only during the day, and if you are always home.
- Using coyote snares. These will catch either dogs or coyotes. Check the legality of snares, in your area, before using.
- Keeping a dog of your own, who will make a commotion if another dog comes by. This can alert you, as well as distract the trespassing dog.
- Having a gun. Even a pellet gun can drive off an attacking dog. However, a dog running among sheep is not an easy target.
- "Live traps" (cages) are useful when trapping dogs, allowing harmless animals to be set free. These are of little value with coyotes, who are too wily to be caught.

DOG LAWS

Your county agricultural agent should be able to tell you the county's dog laws, or better yet, to give you a copy of the county or state laws. You may find that they are strict and well spelled out, but lack enforcement.

Sheep owners should know the law, and work for more adequate dog control legislation if it is needed. These laws should permit the elimination of any trespassing dog that is molesting livestock. Washington state law requires that the owner of the dog destroy it, or be guilty of a misdemeanor. The law should also require payment by the owner (if known) of the offending dog, for both damage and deaths to livestock.

The law should require prompt action by enforcement agencies, and should require payment by the county or state to the livestock owner for losses incurred to his or her sheep (or other stock) from any unidentified dog or dogs not apprehended by the proper officials. While the amount paid is seldom adequate payment for the loss (as in Washington State), enough claims of damage will make the officials more strict with their dog laws.

GATES

Any gate the sheep will be using regularly should be a wide one. Narrow gates and narrow doorways are dangerous when ewes are pregnant, for being crowded and bumped can cause abortions.

It is far easier to lead sheep through a gate than to drive them. If they are in the habit of getting a bit of grain now and then, even at times of the year when it is not completely necessary, you have a way of controlling them. Just rattle the grain in a pan and they will follow you.

There are gate-hardware kits in supply catalogs, having all the materials for the gate with the exception of about five boards.

With wide gates, you will need a heavy gatepost, and if it is a tall gate, you can suspend the latch end of the gate to the post with a cable so the gate will not sag.

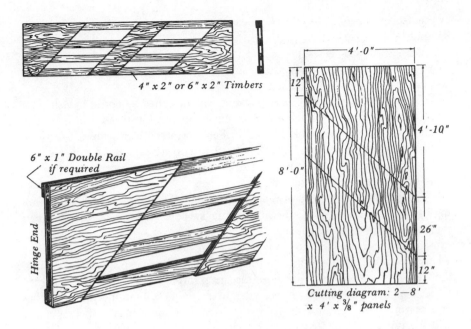

4" x 2" or 6" x 2" Timbers

6" x 1" Double Rail if required

Hinge End

Cutting diagram: 2—8' x 4' x ⅜" panels

Two 8'x4'x3/8" sheets of waterproof plywood are cut as shown at right to make this tough, long-lived gate. Rust-resistant nails or screws are used to fasten plywood to 2"x4" or 2"x6" timbers.

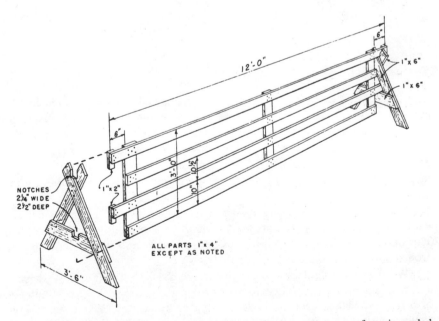

NOTCHES
2¼" WIDE
2½" DEEP

ALL PARTS 1"x 4"
EXCEPT AS NOTED

This portable sheep fence is easy to move and handy wherever a temporary fence is needed. (Utah State University)

If you use a turnbuckle in the cable, it can then be tightened as necessary to avoid sagging.

With heavy wide gates, the use of an old wheel from a discarded wheelbarrow to support the latch end will make the gate roll open and closed easily, and the gate will not sag from its own weight.

The upper illustration shows a plywood gate designed by Forest Industries of Canada, which uses two sheets of plywood cut diagonally into three pieces, to make a gate any width from 8 feet to 16 feet. It should be assembled with rust-resistant nails or 3/8-inch diameter galvanized bolts. The exterior plywood gate is not subject to joint shrinkage and transverse stresses, and may be expected to last twenty years or more if painted with wood preservatives.

For gates in cross fences, it is a good idea to anticipate "forward creep grazing" and make allowance for adjustable openings where the lambs are allowed to go into new pasture ahead of the ewes.

There are several advantages in placing gates between pastures in the corner where the cross fence joins the outer fence. A braced-line post can serve as one of the fence posts. And it is easier to get the sheep through a gate that is in a corner rather than in the middle of a fence. When you need to pen the sheep for shearing or hoof trimming, or catching them for any reason, you can set up a temporary corral with just two movable fence sections, and drive or lead the sheep through the corner gate into this enclosure. With one movable fence section you can set up a chute, with the outer fence as one side of it, for loading sheep into a vehicle.

Inexpensive hinges for narrow people-gates can be rectangles cut from old auto tires. They are flexible and easily replaced when they do wear out. The convex rubber, mounted properly, also keeps the gate closed.

ELECTRIC FENCING

My early experience with electric fencing for sheep and goats was not too encouraging, but this is understandable because we were uninformed amateurs using too few wires and a second-hand fence charger of unknown output, on wooded hilly pasture. After hanging chains on the goats' collars so they wouldn't go over, and wire antennae on their horns so they wouldn't go under, and still having problems, we just gave up and gradually fenced all the property with sturdy woven wire, barbed on top.

More recently, I have found a new and most successful use for electrified fence, with 3-foot-high chicken wire around a stand of corn to protect it from raccoons. This was made more effective with grounded 2-foot chicken wire laid flat on the soil around the outside.

When considering electric fencing for sheep, check with your county extension agent, because ordinances differ from one area to another regarding the installation of the fences and the legal types of chargers and control boxes. Be sure that the controller you use is not a shock or fire hazard, and is approved by authorized authorities.

It is important to use one of the new high-voltage low-impedance fence ener-

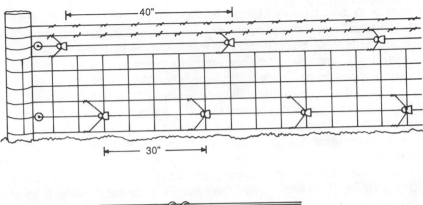

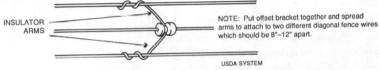

INSULATOR
ARMS

NOTE: Put offset bracket together and spread arms to attach to two different diagonal fence wires which should be 8"–12" apart.

USDA SYSTEM

There is also a two-wire offset dog and coyote control system that can be installed to supplement an existing woven wire fence that has the usual barbed wire strands at the top.

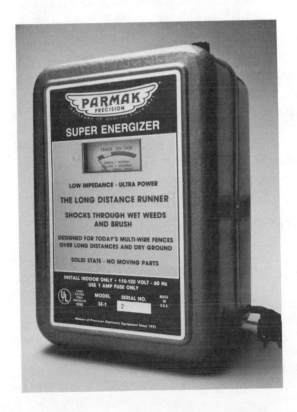

Parmak fence charger.

gizers with any type of electric fencing. The use of "weedburner"- or "bullwhacker"-type chargers is now outmoded. They can damage electroplastic fencing, are easily grounded out by grass or weeds, and can set pasture fires. They do not have the holding power of the new generation energizers, and will not charge as much fencing as the modern ones. Until recently, most low-impedance energizers were imported from New Zealand and were expensive. Now, one of the oldest manufacturers of chargers in this country is making a Parmak-brand Super Energizer with a built-in voltage meter that provides a constant readout of the fence's condition. My veterinarian has one, a Model SE-1, and says it is about three times more powerful than his previous New Zealand import, and was less than half the price.

On his recommendation, I got one and now I can see that any previous electric fence problems I had were really "fence-charger problems," due to inadequate charger.

This SE-1 is approved by Underwriters Laboratories (UL) in the U.S. and also by the Canadian Standards Association (CSA). There is also a two-wire offset dog and coyote control system that can be installed to supplement an existing woven wire fence that has the usual barbed wire strands at the top.

An "outrigger" wire can be added to a woven wire fence having wooden posts. Take fiberglass fence stays, cut them into 1-foot lengths, and sharpen the ends. Drive these into the posts, one per post, at an angle. Slip on an insulator, and run one electrified wire. The height of the wire would depend on whether you were preventing dogs from digging under, large dogs from jumping over, or small dogs from going through a wide mesh.

PORTABLE FLEXIBLE ELECTRIC FENCING

There are several types of portable electric fencing. Most are made of polywire, which is polyethelene and wire filaments woven together into strands or ribbons, often called electroplastic. These are light, flexible, and carry a charge well. Black polywire is the longest-lasting color, but not too popular as it is more difficult for either people or animals to see. White and orange are the most used colors. There is even an extra-wide ribbon type of polywire that can be seen from quite a distance. The "netting" portable fence and the "reel" type are easily available in all parts of the country.

One well-known brand of portable fence is Electronet, made in heights of 22 inches, 33 inches, and 42 inches, with posts already built into the roll and semi-rigid yellow plastic verticals spaced every 12 inches to allow the installation over quite severe changes in surface terrain. The fence can rise or fall as much as 10 inches between the 13-foot post placement. For extra support, many users install several extra fiberglass rods along each roll of fencing.

This "instant"-type portable fencing is intended mainly as a temporary fence. While it can be used to bolster existing permanent fencing that has become run down and really needs mending, this electrified plastic-covered wire is at its best when used as cross-fencing to rotate pastures, for an additional predator-control around a night corral, for temporarily fencing off an orchard during a time of excess fruit on the ground, for separating lambs for weaning, or as part of a garden protection system. It could allow the grazing of a lawn from time to time by a ewe with twins or triplets, giving unusually fresh grass to a select few.

Booklets are available on both portable and permanent power-fencing systems (see Sources). Most fencing manufacturers have pamphlets on the use and installation of their products, and are most anxious to help.

It is important to use the type of fence charger specified in the fencing instructions. It is seldom necessary to use the exact brand-name charger recommended, but the right type is important.

TRICKS

There are ways to make electric fences more effective, especially in places of known predator menace. In extra-dry weather, calcium chloride can be sprinkled on the ground along the outside of the fence to attract moisture from the soil.

Another trick is "baiting" the fence in special places. This is very effective with long-furred animals, as it usually zaps the nose. A can of sardines is wired to the fence, about 20 inches off the ground, and punctured in several places so that it drips sardine juice down onto the lower wires. The animal will be attracted by the scent, and will put its nose against the can or the wires that are wet with sardine juice.

To make the electric fence more effective in dry conditions, we lay a 2-foot-wide strip of chicken wire along the outside of the fence, connected to the ground terminal of the charger and casually staked down so it doesn't buckle. Make sure the chicken wire does not contact the hot wire of the fence.

FENCE CHARGER POWER

There are two methods of evaluating output from a fence controller. One is rated Millicoulombs and is the method recognized by all North American testing laboratories, including UL and CSA. Millicoulombs is a factor of time and current in amperes, which is the predetermined time in milliseconds and the amount of current applied to fence wire by the charger.

Many imported brands refer to joules of output. Joules uses watts rather than amperes to calculate the output of a fence controller. Regardless of the method used, amperage is the determining factor in the "power" of the charger. The UL and CSA use amperage, as it is a more direct and positive method of determining fence safety. Due to the voltage output of modern controllers, the amperage and duration of the spark are closely controlled for the purpose of safety.

PERMANENT HIGH-TENSILE ELECTRIC FENCING

If you wish to design and install your own permanent high-power fencing, there are companies such as Premier and Gallagher Snell that carry a full line of all the supplies needed, such as insulators, in-line tensioners and tension springs, twist-links, and even steel-post insulators.

Four- and five-wire power fences, just over 3 feet, 3 inches high, are in use all over New Zealand for the effective control and protection of sheep on a wide vari-

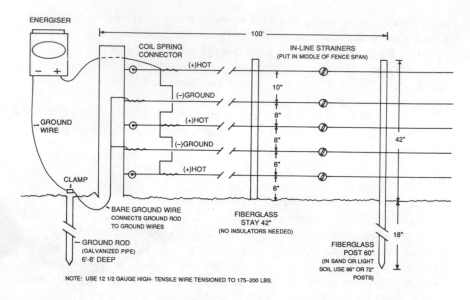

Five-wire high-tensile electric fence system.

ety of land types. Fence experts in New Zealand consider that six wires, with alternative live and earth (ground) wires, are most effective for sheep. There is almost unanimous agreement that 12 1/2-gauge wire is best for permanent high-power fences.

There is an "insultimber" post system that will not need insulators, as the posts are milled from Australian ironbark trees, wood of high density that will not conduct electricity even when wet. These posts need no treatment, will not rot, and are impervious to insects. Few posts are needed, for droppers (a stay that is not driven into the earth) and tie-downs make it possible to space on 50- and 150-foot centers, unless the terrain is very rough.

A USDA-recommended fence system of five high-tensile wires is said to be effective against coyotes, dogs, and coy dogs, and is charged by a low-impedance charger, which maintains high line voltage and is not badly affected by grass and weeds. Because the charged pulse lasts for such a brief period of time, the fence poses little danger. (Chargers are available in electric, battery, or solar-powered units; see Sources.)

CHAPTER 4

Feeds and Feeding

RAISING SHEEP is an efficient way to convert grass into food and clothing for humans, but pasture alone is seldom adequate to feed sheep twelve months of the year, making feeding of supplements (grain and hay) quite necessary.

Poor feeding of ewes results in reduced fertility, poor nursing ability, fewer multiple births, decreased wool production, a higher incidence of pregnancy disease, and reduced growth of lambs. An undernourished ewe may also lamb a few days early, and if the lamb has not reached its full size before birth it has less chance of survival. An undersized lamb, born outdoors in bad weather, suffers considerably more loss of body heat because of its smaller size.

Feeding time is a good time to check on your sheep, feel their udders when close to lambing, and note eating habits, which greatly reflect their state of health. Count the sheep, particularly if you have any wooded pasture where one could get snarled up or be down on its back and need help.

FEEDING

The same quantity of feed supplement is not needed at all times of the year. The grain amounts are planned around the reproductive cycle.

GRAIN FEEDING PERIODS

1. Seventeen days before turning ram in (see Chapter 6), give up to 1/2 pound of grain per ewe, starting gradually for the first few days.
2. Up to three to four weeks after mating, give the same amount, tapering off gradually. This may prevent resorption of the fertilized ova.
3. Light grain feeding until the last five weeks of pregnancy.
4. Last five weeks of pregnancy, when ewe should be on a rising plane of nutrition to prevent pregnancy disease, give 1/2 pound or more per ewe.

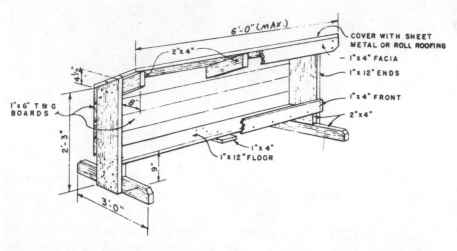

This covered salt box offers protection to the salt from the elements. Plywood can be substituted for the 1″ x 6″ boards. (Utah State University)

5. For six weeks of lactation, ewes with single lambs should have approximately one pound of grain per day, while a ewe with twins should get from 1 1/2 to 2 pounds, plus hay for each. Then, taper off as the lambs eat more grain and hay (in their creep feeder).
6. Withdraw grain and hay from ewes five days before weaning lambs, or start diminishing the quantity ten days before, leaving feed in the creep feeder for the lambs.

In building a barn or shelter, plan for a feed room, or some safe place for storing grain where sheep can't break in and get to it. This can be a closed room or an alcove with a secure gate. Put feed in a garbage can with the lid wired down, or a spring-clamp that will keep the lid on even if it is tipped over. This has the added advantage of being rodent-proof.

Any sudden large amount of grain can paralyze the digestive system of the sheep and cause death from acidosis, impacted rumen, enterotoxemia, or bloat. "Acute indigestion" is not a minor illness for a sheep, which has four stomachs.

REGULAR FEEDING (TIME AND AMOUNT)

Measure the quantity of grain given each day, by using the same container, or number of containers, for each feeding. Sheep do not thrive as well when the size of their portions fluctuates. If they are fed in the evening, it should be at least an hour before dark, for sheep are not like cattle and horses; they do not eat well in the dark and should have time to eat their food before nightfall.

When given regular feedings at an expected time, they are less apt to bolt their feed and choke. Too much variation in feeding time is hard on their stomachs and

Suffolk sheep find this hay and grain feeder a good gathering place. Drawings show how to construct it. Board C can be made of 1″ x 10″ for large ewes, or from 1″ x 8″ for smaller breeds. For use with lambs of small breeds, the dividers B can be made from 1″ x 6″ instead of 1″ x 2″. (Photo from National Suffolk Sheep Association. Drawing from Midwest Plan Service)

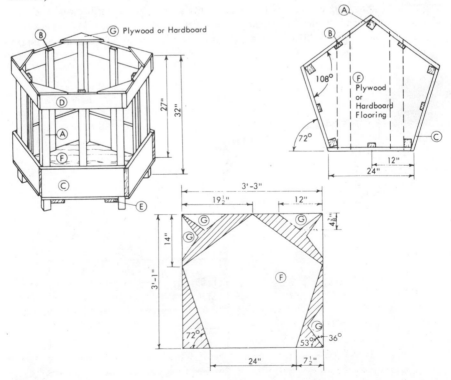

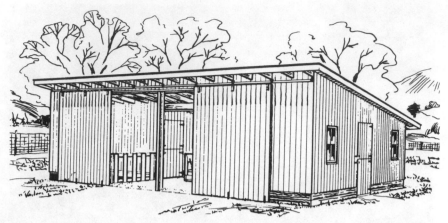

Sheep and Lambing Shed for 30-36 Ewes.

This sheep barn has a grain room, space for use of the hinged lambing pens, and a lamb creep area with a removable creep entry panel so hay can be stored there when the area isn't in use for lambs. The feed storage room is safe and secure from the sheep. Store grain in rodent-proof containers. The building does not have to be heavily insulated, but should be tight enough to prevent drafts. If electricity is available, circuits should be provided for heat lamps when needed for newborn lambs.

The building is 20 feet deep, 32 feet long and provides space for about 30 ewes. The length of the building can be increased in multiples of 8 feet. It needs no foundation and creosoted timbers are used. This is UDSA Plan #5919, available in two pages of blueprints at approximately $1 per page. (USDA plan)

Plan

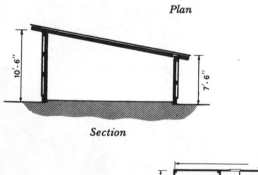

Section

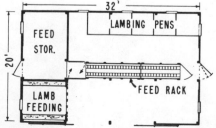

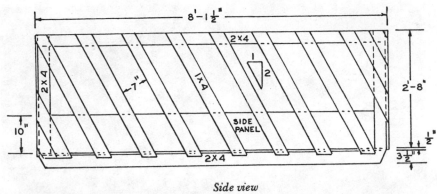

Side view

This is a good hay feeder for the sheep barn shown. These slanted slats discourage the sheep from stepping back with a mouthful of hay and dropping some on the ground.

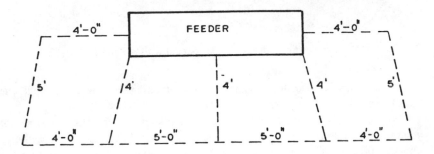

Lambing pens around feeder, up to 8 pens per unit.

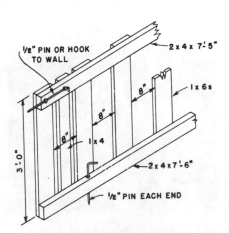

A removable creep fence for lamb feeding area inside barn. (USDA plan)

A pan of grain makes the author a popular person with her sheep. (Joseph Scaylea)

their entire systems. During pregnancy, erratic feeding can trigger toxemia. (See also Ketone Test, in Chapter 7.)

While regular feeding time is important, it also makes a difference what *time* you feed, for pregnant ewes. In some documented trials, regular feeding of ewes about 10:00 a.m. helped reduce the night-to-early-dawn lambing incidences. Other recent tests suggest late afternoon feeding, shifting even later in the day as lambing approaches. Either feeding schedule concentrated lambing primarily into daylight hours.

Rumen papillae on lamb that was on diet of pelleted alfalfa (left) and on diet of chopped hay (right). (USDA Ruminant Nutrition Laboratory)

FEEDS

GRAIN

Whole grains, with the exception of barley which can lead to metabolic disturbances in pregnant ewes, are better for sheep than crushed grain. Rolled oats, for instance, has so much powdery substance that it can cause excessive sneezing, leading to prolapse in heavy pregnant ewes, and breathing problems in lambs. Unprocessed corn and wheat still contain the valuable germ rich in vitamin E that ewes need to help protect their lambs from white muscle disease.

Whole grains promote a more healthy rumen. A diet consisting mainly of pelleted feed causes the papillae of the rumen to lump together and become inflamed. This traps debris and causes more inflammation. Whole grain, on the other hand, promotes a healthy rumen wall, where the feed gives better conversion to growth.

FEED CHANGES

A sheep's stomach can adjust to a great variety of feed, providing changes are made *gradually*. A sudden change of ration, such as sudden access to excess food, can cause death. The rumen has a mixed bacterial content with the ability to adapt to the nature of the diet. Sheep who are fed only grain will not be able to adapt to a sudden change to hay, for the rumen will be so geared to the handling of concentrated starch and protein that the bacteria which digest cellulose will be present in too small a number to function, and the sheep will go off their feed and suffer. A gradual change from grain to hay gives those cellulose-handling flora a chance to

An old hot water tank, cut in half lengthwise, makes a good outdoor feeder for grain. (Charles R. Pearson)

multiply. A disturbance of the rumen by an abrupt change of diet will leave the sheep open to infections and disease, by interfering with the synthesis of A and B vitamins, vitamin A in particular being the anti-infection vitamin. A good rule of thumb would be to change feed no faster than 10 percent per day.

HAY

One reason alfalfa hay is such a superior feed for sheep is its content of nine vitamins, especially vitamin A that is so lacking in winter pasture grass. The greener the hay, the higher the vitamin content. It is also high in calcium, magnesium, phosphorus, iron, and potassium. Protein content is from 12 percent to 20 percent depending on what stage it is cut (highest protein when cut in the bud) and on its subsequent storage. Alfalfa got its name from an Arabic word meaning "best fodder," which is most appropriate.

Hay should be stored in the darkest part of the barn to preserve its vitamin A, which is depleted by exposure to sunlight. Careful storage is necessary to avoid weather damage and nutrient loss. Exposure of hay to rain can not only leach out its minerals, but can also result in moldy hay, one cause of abortion in ewes.

The lower the hay quality, the more of it you will have to feed. Lots of heavy stems in the hay will mean more that the sheep will not eat. A certain amount of hay is always discarded, some pulled out onto the ground and wasted (pile this in your garden twice a year) and some uneaten stems (save these clean ones out of the feed rack for clean bedding for lamb pens). If you buy two different kinds or grades of hay, save the best for the pregnant ewes. Late in pregnancy, hay must be of high quality, as the growth of the lamb will crowd the ewe's stomach and leave little room for bulky low-nutritive feed.

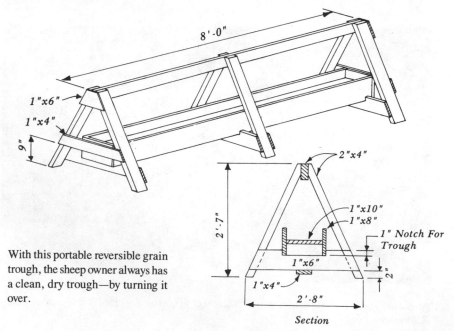

With this portable reversible grain trough, the sheep owner always has a clean, dry trough—by turning it over.

Section

EXTRAS

Windfall apples, gathered and set aside out of the rain, can be a welcome addition to the winter diet, but in limited quantities. Sheep love apples, even preferring the over-ripe and spoiled ones, and a few apples a day will add needed vitamins. An excess of apple seeds, however, especially the green seeds, is toxic.

Fresh pomace from apple cider making is good feed for sheep, in small quantities, if you have not sprayed your apples. Fermented pulp is not harmful if fed sparingly, but decomposed pulp is toxic.

Molasses is another treat for sheep, and a good source of minerals. As its sugar enters the bloodstream quickly, it is of value to ewes, late in pregnancy, to prevent toxemia—but not in excess.

Discarded produce from the grocery store is another treat. Lettuce, cabbage, broccoli, celery, and various fruits, past their prime for human consumption, are often available at the local store. Fed sparingly, or regularly in measured quantities, they are a good addition to the diet.

WATER AND SALT

Hot summer pasture has very little moisture in it, so sheep need more water, not just because of the heat, but because their feed contains less liquid. To cope with the heat, sheep lose more moisture through their skin, which adds to the need for ample water. Providing shade helps keep down their moisture loss, but they still need clean fresh water. Ewes with lambs need access to plenty of clean water in order to make milk.

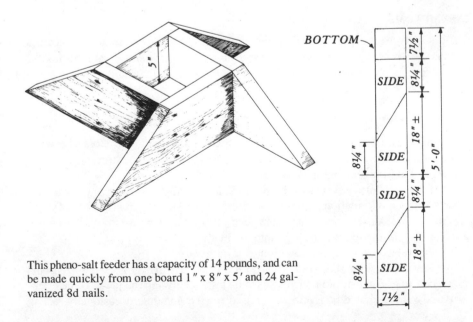

This pheno-salt feeder has a capacity of 14 pounds, and can be made quickly from one board 1″ x 8″ x 5′ and 24 galvanized 8d nails.

Salt is another year-round necessity for good health. When sheep have been deprived of salt for any length of time and then get access to it, they may overindulge and suffer salt poisoning. (Treatment: access to plenty of fresh water.) Avoid this illness by keeping salt available at all times. With the advent of new efficient wormers, the addition of low-level amounts of phenothiazine to salt went out of favor for a while, but is now recognized as being beneficial, in addition to regular worming practice. The use of mineralized salt, especially mixes containing selenium, is recommended, with phenothiazine added to the salt. Just make sure it is a sheep mineral-salt, as minerals for cattle can contain toxic levels of copper.

Regular access to salt is said to be useful, along with roughage, to prevent bloat, which is one of the most serious digestive upsets.

BLOAT

Bloat is an excessive accumulation of gas and/or foamy material in the rumen. Severe cases can be fatal in as little as two hours if not treated.

Too much of almost any feed can cause bloat, but overconsumption of wet clover, grain, orchard fruit, or lush pasture is the most common cause. Legume pastures, such as very leafy alfalfa and clover, are even more dangerous than grass.

When changed from sparse to lush pasture, sheep may gorge themselves *unless* given a feeding of dry hay prior to turning out on the new pasture. Sheep seldom bloat when they are getting dry hay with their pasture. The coarse feed is thought to stimulate the belching mechanism, as well as keeping the green feed from making a compact mass. Some sheep seem more prone to bloat than others, possibly due to a faulty belching mechanism.

SYMPTOMS

Enlargement of the rumen on the left flank is a sign of bloat. The animal breathes hard, grinds its teeth because of abdominal gas pain, sometimes salivates profusely, and stops eating. When the sheep falls to the ground, death usually follows, probably from suffocation.

TREATMENT

Simple treatments are more likely to succeed if bloat is noticed before it becomes too severe. There are commercial preparations on the market for treating bloat, which are good to have on hand in case they are needed.

If bloat is not so severe as to have caused a breathing disturbance, you can prevent further gas formation by giving two tablespoons of baking soda in a cup of warm water, using a dose syringe. Do not raise the sheep's nose above its eye level when doing this, or the mixture may go into the lungs.

Repeat the dosage in thirty minutes if necessary.

You can place the sheep in a sitting position, and massage the abdomen to encourage belching. Another old remedy was to tie a stick between the jaws. To be effective, the sheep should be forcibly exercised when doing this, occasionally raising its front feet off the ground to encourage the release of gas. This is of little value if the bloat is of the foamy type. Foam, which cannot be belched up, can be helped by another one-half cup of vegetable oil (peanut oil or corn oil are recommended) given by mouth if the ewe is still able to breathe and swallow normally.

A 3/8-inch or 1/2-inch rubber tubing (small siphon tube) down the throat into the stomach can release gas, unless there is too much foam. If you use the finger test to be sure it is not in the lungs,* you can break up the foam somewhat by pouring 1/3 to 1/2 cup of vegetable oil into the tube with a funnel. In an emergency, the rumen can be punctured, preferably by a vet, with proper equipment to relieve both foam and gas, and to treat to prevent infection.

POISONOUS PLANTS

Red maple leaves	cause kidney damage
Acorns	cause kidney damage
Rhododendron	
Mountain Laurel	these glossy leaf plants are all toxic
Azalea	
Yew needles	extremely toxic
Apple seeds	especially the green seeds, poisonous
Nightshade	poisonous to animals and humans (goats like it)
Skunk cabbage	causes birth defects
Wild tobacco stalks	causes birth defects
Potato sprouts	causes birth defects
Death Camas	poisonous, especially in the spring
Horsebrush	poisonous, especially in the spring
Lupine	poisonous, especially in summer and fall
Milkweed	poisonous, especially in summer
Chokeberry	poisonous, especially in spring
Loco	poisonous
Halogeton	poisonous
Water hemlock	poisonous, but not palatable
Larkspur	poisonous at some season; dangerous (sheep like it)
Tansy ragwort	poisonous

Check with local authorities, such as your county extension agent, as to what plants in your particular area are poisonous. With any new pasture, walk around it and note any unusual or unfamiliar plants. Find out what these are, and if toxic, they should be eradicated before pasturing sheep there.

If there are plants you cannot identify, you can send several fresh whole plants (stem, leaves, flowers, seeds) to the state agricultural college, wrapping them in several layers of newspaper. They should be able to identify them and advise as to toxicity.

---

*Finger test to determine if a tube is in the stomach where it belongs or the lungs: Wet a finger and hold it in front of the protruding end of tube. If you feel cool air, like breathing, you are in the lungs. Pull out the tube and try again.
After pouring oil down the tube, withdraw it fairly quickly, to avoid dribbling any oil at the entrance of the lungs.

When plant poisoning is suspected, it is important to call your veterinarian promptly. Knowing what is the cause of the poisoning can make effective treatment more possible, of course. Keep the sick animal sheltered from heat and cold, and allow it to eat only its normal safe feed.

In most instances, animals do not happily eat toxic plants. When overly hungry, however, and better food is not available, they may eat anything at hand. If they lack sufficient water for an extended period, this can cause them to reduce their feed intake. Then, when they suddenly get ample water, their appetites increase greatly, and they may devour almost anything that they can get. You can see how important it is to have water and to feed at regular times, in needed quantities.

Overgrazing of pastures, which means shortage of grass, can cause sheep to eat plants that they would otherwise avoid. Better to keep fewer sheep, well fed and healthy, than to keep more than your pasture and pocketbook can sustain.

TOXIC SUBSTANCES THAT CAN CAUSE ILLNESS OR DEATH

- Waste motor oil, disposed of carelessly.
- Old crankcase oil (high lead content).
- Old radiator coolant or antifreeze (sweet and attractive to sheep).
- Orchard spray dripped onto the grass.
- Weed spray (some have a salty taste).
- Most sheep dips.
- Old pesticide or herbicide containers, filled with rainwater.
- Old auto batteries (sheep like the salty lead oxide taste).
- Salt. Sheep require salt for health. When deprived, then allowed free access, they may ingest large quantities, causing salt poisoning. Symptoms are trembling and leg weakness, nervous symptoms, and great thirst.
- Commercial fertilizer. It has a high nitrite content, and in the rumen of sheep is converted to nitrite, causing death. Be careful not to spill any fertilizer where sheep can eat it, and store the bags away from the sheep. They may nibble on empty bags. Several rainfalls are needed after fertilizing a field, and it still may not be safe unless the pasture grass is supplemented with grain and hay. Symptoms are weakness, rapid open-mouth breathing, and convulsions. For a home remedy, use vinegar, one cup per ewe, as a drench. A veterinarian will have a more certain treatment, if started soon enough.
- Cow supplements containing copper. Some cattle mineral-protein blocks contain lethal levels of copper, for sheep. Some mixed rations intended for cattle may also have copper and should not be used for sheep.
- "Empty" lead buckets, filled with rain water. (Rain water is soft and readily dissolves enough lead to kill a sheep that drinks from it.)

CHAPTER 5

The Ram

THE OLD SAYING that the "ram is half the flock" is still the best rule of thumb for selection of a flock sire. Even though scientists have discovered that certain inheritable traits are not transmitted equally between the sire and the ewe, the choice of sire will most rapidly change the character of your flock, good or bad, so inspect him well before buying. The following points should be kept in mind when looking at a ram for possible purchase:

- Good size, deep wide body, heavy muscular neck.
- Well-developed sex organs, scrotal circumference 12 inches or more (for 150-pound yearling).
- No scrotal mange, no hernia. Turn the ram up to inspect.
- Insist on a negative ELISA test for ram epididymitis.
- Good feet. Bad feet can render a ram useless.
- Good eyes. Watch for pinkeye or any sign of eye damage.
- Good teeth, well aligned with upper jaw.
- Head not too large, to avoid hereditary problems in lambing.
- Full hindquarters, so his lambs will be good meat animals.
- Good fleece, of a type to suit your wool market.
- One of twins or triplets, to influence prolificacy of his daughters.
- Notice the general health of the rest of the flock.
- Lack of external parasites. Their presence signifies negligence.

Ordinarily the best investment is a well-grown two-year-old, a twin or a triplet. Being a twin will in no way affect the twinning of the ewes he breeds. This is controlled by the number of eggs the ewe drops to be fertilized, which is influenced by genetics and encouraged by flushing. However, the lambs they have will inherit both the ewe's and the ram's twinning capabilities, and this will show up in the following generations. The ram greatly influences conformation and fleece type.

If he is a lamb, use him sparingly for breeding. One way to conserve his energy is to separate him from the ewes for several hours during the day, at which time he can be fed and watered and allowed to rest.

One good ram can handle twenty-five to thirty ewes. On a small flock where the ram gets good feed, you can expect about six years of use from him. On open range, where there may be over-use with more ewes per ram, fighting with other rams, and little supplemental feed, rams get run down during the breeding season from eating so little and chasing the females. They then succumb to diseases because of their low resistance. If you are buying a new ram, do this long enough before breeding season so that he becomes acclimated to his new home. Keep him separate on good feed and pasture until breeding time, but if you are going to feed him a different ration than he had previously, be sure to change gradually. Use good judgement in feeding. Excess weight results in a lowering of potency and efficiency, so keep him in good condition, but not fat.

During the breeding season, feed the ram about one pound of grain per day, so that if he is too intent on the ewes to graze properly, he is still well nourished. After separating him from the bred ewes, a maintenance ration of one-half pound of grain per day, plus hay as necessary during the winter, should carry him through until good pasture is available again.

Provide a cool shady place for him in the heat of summer. An elevated body temperature, whether from heat or even an infection, can cause infertility. Semen quality is affected at 80 degrees, and seriously damaged at 90 degrees air temperature. Several hours at that temperature may leave him infertile for weeks, and cancel your plans for early lambing. If your climate is very hot in the summer, shear his scrotum just before the hot weather, and run him with the ewes in the evening, at night, and in the early morning, but keep him penned in a cool place during the heat of the afternoon, with fresh water. (High humidity and temperature can also decrease his sexual urge or instinct.)

August is generally the beginning of breeding season for early (January) lambing. You can wait until later to turn the ram in with the ewes, if you want to start lambing later in the spring. Gestation is five months (148 to 152 days). There is a computer program available (see Sources) to set up your whole year's plans (including vaccinations) around your desired time of lambing.

Ewes are in heat about twenty-eight hours, with about sixteen to seventeen days between cycles, so fifty-one to sixty days with the ram should mate all the ewes, even the yearlings who are sometimes late in coming in heat.

A sense of smell greatly determines a ram's awareness of estrus in the ewes. A study of sex drive in rams, done at University College in Aberystwyth, Wales, found that some breeds of rams have keener olfactory development than others, and are able to detect early estrus in ewes that goes unnoticed by other breeds. Those with the "best noses" for it were singled out as Kerry Hill, Hampshire, and Suffolk rams, in that order.

EFFECT OF THE RAM ON EWES

The presence of the ram, especially the smell of the ram, has a great effect on estrual activity of the ewes. This stimulus is not as pronounced when the ram is constantly with the ewes as it is when he is placed in an adjoining pasture about two weeks ahead of the time when you would like the breeding season to start. Owners of large flocks often use a vasectomized ram, turned in with the ewes about three

A newly sheared Suffolk ram, owned by E. William Hess, Barboursville Farms.

weeks prior to the scheduled breeding, in order to stimulate the onset of estrus in the flock.

There is actually a wide range of sexual performance among rams. Anyone who has had more than one ram at a time will be conscious of the "social" differences— one ram must always assert dominance. Any time the rams are separated for a period, there is the inevitable fighting and head-butting until the "boss" is decided.

It has been documented that the mating success of dominant rams does far exceed that of the subordinate ones. This in itself can cause problems, since aggressive potential and ram fertility are not necessarily related. If the dominant ram is infertile for any reason, then flock conception rates can suffer.

RAM-MARKING HARNESS

To keep track of the ewes who are bred, and when, a "marking harness" is available in many sheep supply catalogs for use on your ram. It has a holder on his chest for a marking crayon. Each ewe is marked with the color of the crayon the day he breeds her. Inspect the ewes each day and keep track of the dates so you will know when to expect each one to lamb. Use one color for the first sixteen days he is with the ewes, then change color for the next sixteen days, and again for the next. If many ewes are being re-marked, it means they were not bred the previous times he tried to breed them, as they are still coming into heat, so you many have a sterile ram.

If the weather was extremely hot just before or after you turned him in, you can blame it on the heat. But to be safe, you should get another ram in with them, in case your ram's infertility is not just a temporary one caused by hot weather.

PAINTED BRISKET

Instead of a purchased harness, you can daub marking paint on the ram's brisket (lower chest). Mix it into a paste with lubricating oil, or even vegetable shortening, using only paints that will wash out of the fleece. Suggested colors to use: Venetian red, lamp black, yellow ochre.

RAISING YOUR OWN RAM

If you are raising market lambs for meat, you might try a system called "recurrent selection of ram lambs," which consists of keeping the fastest-gaining ram lambs sired by the fastest-gaining ram lambs. No, this is not a misprint. Recurrent selection of ram lambs is a way of improving the potential for fast growth in your lamb crop. It involves changing rams fairly frequently, and leaves you the problem of disposing of a two- or three-year-old ram. If he is a good one, you can probably sell him as a ram, trade him to another sheepraiser—or see the chapter on "Muttonburger."

One advantage of raising your own ram is that you see what he looks like at what would be market age, if he were sold for meat. The older a ram gets, the less you can tell about how he looked as a lamb, or how his offspring will look when they are market age.

The way ram lambs are raised can have some effect on their future sexual performance. Studies have shown that rams raised from weaning in an all-male group will show lower levels of sexual performance in later life. Some will actually show no sexual interest in receptive females.

When you are raising a lamb for a breeding ram, do not pet him or handle him unnecessarily. Do not let children play with him, even when he is small. He will be more prone to butting and becoming dangerous if he is familiar with you than if he is shy, or even a little afraid of humans.

THE "BATTERING" RAM

In actual cases, the butting ram is not very funny, and can inflict serious, sometimes permanently crippling injuries. Keep children away from a ram. He may hurt them badly, and they can make him playful and dangerous. Never pet him on top of his head; this encourages him to butt.

Leading a ram with one hand under his chin will keep him from getting his head down into butting position. A ram butts from the top of his head, not from his forehead. His head is held so low that as he charges you, he does not see forward well enough to swerve suddenly. A quick step to the right or left will let you avoid the collision.

If you have a ram who already butts at you, try the water cure: a half-pail of water in his face when he comes at you. After a few dousings, a water pistol or dose syringe of water in his face usually suffices to reinforce the training. Adding a bit of vinegar to the water makes it even more of a deterrent.

A dangerous ram who is very valuable can be hooded so that he can only see a bit downward and backward. He must then be kept apart from other rams, as he is quite helpless.

Strange rams will fight when put together. Well-acquainted ones will, too, if they've been separated for a while. They back up and charge at each other with their heads down. Two strong rams, who are both very determined, will keep at it until their heads are bleeding, and one finally staggers to his knees and has a hard time getting up. Rams will occasionally kill one another. (Never pen a smaller, younger ram with a large dominant one.) Once they have determined which one is boss, they may playfully butt, but will fight no big battles unless they are separated for a time.

To prevent fighting, and the possibility of one being badly injured, you can put them together in a small pen for a few days at first. In a confined area, they can't back up far enough to do any damage.

If no pen is available, they can be "hoppled," "yoked," or "clogged"—all old European practices.

Hoppling a ram (the modern term would be *hobbling*) was an old system of fastening the ends of a broad leather strap to a fore and a hind leg, just above the pastern joints, leaving the legs at about the natural distance apart. It discourages rams from butting each other, or people, because they are unable to charge from any distance, and they cannot hit very hard if they can't get a run. They may stand close and push each other around, but will do nothing drastic. Hoppling also keeps them from jumping the fence, which rams will sometimes do if ewes are in the adjoining pasture.

Author is shown being chased by her least favorite ram. (Washington State Department of Commerce and Economic Development)

Yoking is fastening two rams together, two or three feet apart, by bows or straps round their necks, fastened to a light timber like a 2-inch by 3-inch piece of lumber.

Either of these methods will necessitate watching to be sure that the rams do not become entangled.

In *clogging*, you fasten a piece of wood to one foreleg by a leather strap. This will slow down and discourage both fence jumping and fighting. Close watching is not necessary.

RAM EPIDIDYMITIS

Scientists, researchers, and drug companies have all been working to find solutions to the ailments that can interfere with ram virility and ewe fecundity, and the last ten years have seen several important findings.

There is recently much more awareness of ram epididymitis, a disease caused by one of several different organisms which damages sperm-producing tissues. The infection is well under way and contagious before the external symptoms show up in physical examination. Symptoms can include swelling of the epididymis (located at the base of the testis) and the presence of hard lumpy tissue, showing that the disease is far advanced. In some cases the ewe may become infected, resulting in abortions, stillbirths, and weak lambs. It is mainly contagious from ram to ram, but can be transmitted through a ewe who has been serviced by a diseased ram. Vaccination has not yet proved highly successful, and can interfere with diagnostic tests.

Accurate diagnostic procedures are now available so that diseased rams can be readily identified and culled. When the only indication of the disease is seminal white blood cells, before the disease becomes clinical, high levels of antibiotics such as tetracycline and streptomycin can be effective. This could be a way of saving a valuable animal, but would require isolation and extensive monitoring.

A new blood test is highly sensitive and specific for identifying ram epididymitis—the Enzyme-linked Immunosorbent Assay (ELISA) test, which became available in 1986 from the Western Slope Diagnostic Laboratory, 425 29th Road, Grand Junction, CO 81501 and from the Department of EPM, College of Veterinary Medicine, University of California, Davis, CA 95616. Its cost is moderate and depends on the number of samples sent, so a veterinarian can collect and freeze samples from several flocks to submit together.

The state of Colorado now requires that any ram sold for breeding purposes in Colorado must have a negative ELISA test. This requirement in all states would greatly protect sheep buyers. If not required by law, it should still be demanded by the buyer.

CHAPTER 6

Flushing and Breeding

THE OPTIMAL TIME for lambing varies greatly among geographical areas. The desired lambing time may depend on the availability of pasture, local weather conditions, labor/time restraints, targeted lamb markets, etc. Choose your lambing time to fit your priorities, and plan to breed about five months before you want lambs.

When the cost of hay or grain is a consideration, lambing should be timed to take advantage of new pasture growth. It takes the rumen of newborn lambs about six weeks to develop, or a little less if creep feeding starts as early as ten days old. So you could plan for the lambs to be about five to six weeks old about the time of the first good early growth of pasture. Shepherds living in the Midwest with moderate winters and hot summer temperatures often lamb in the autumn or early winter to maximize weight gains, knowing that lambs experience very poor weight gain in hot temperatures. Those in the northern states often begin lambing in March or April in order to avoid the severe sub-zero temperatures, while those in temperate coastal climates may let the rams run with the ewes the year round and let nature take its course, if they have no target date for market lambs. What constitutes "early" or "late" lambing will depend on your climate.

Here are the advantages of both early and late lambing:

EARLY LAMBING

1. There are fewer parasites on the early grass pasture.
2. Ewe lambs born early are more apt to breed as lambs.
3. You can sell early lambs by Easter, if creep fed, and get a better price for early meat lambs.
4. You can have all lambs born by the time of the best of the spring grass.
5. There are fewer problems with flies at docking and castrating.

LATE LAMBING

1. It is easy to shear ewes before lambing.
2. It avoids lambing danger in severe weather.
3. Mild weather means fewer chilled lambs.

4. Ewes can lamb out on the pasture.
5. Less grain is required for lambs, since you have lots of pasture.

GET READY TO START BREEDING

Worm your ewes, and trim away any wool tags from around the tail. Trim their feet, for they will be carrying extra weight during pregnancy and it is important that their feet be in good condition. Worm the ram, too, and check all for ticks. If you eliminate ticks before lambing, none will get on the lambs and you will not have to treat for ticks again.

At seventeen days before you want to start breeding, put your ram in a pasture adjacent to the ewes, with a good fence between them. Australian research has determined that the sound and smell of the ram will bring ewes into heat earlier. They obtained a similar reaction just by fastening a ram-scented pad to the ewe's nose. (I feel tempted to compare this to human use of after-shave lotion and perfume.)

Some large flock-owners have initiated the use of a vasectomized (sterilized) ram to stimulate the onset of estrus in the flock. Since it always seems that the male lambs make the best pets, this is one way you can keep a pet without feeding a non-productive wether!

Never pen your ram(s) next to the ewes before this "sensitizing" period just prior to breeding. Remember, "absence makes the heart grow fonder." It is the sudden contact with the rams that excites the females.

VACCINES

One vaccine that is most important to both ewes and their lambs is Covexin-8. Ewes need it twice the first year—the "primer" shot can be given as early as breeding time or as late as six to eight weeks prior to lambing, with the "booster" shot given two weeks prior to the calendar lambing date for the flock. For subsequent lambings they require only the booster given two weeks prior to lambing. This protects ewes, and lambs up to about ten weeks of age, against all the clostridial diseases including tetanus.

Other important vaccines:

1. Two-way abortion vaccine (EAE-Vibro), from two weeks prior to breeding up to one month after breeding. This protects against abortions caused by Chlamydia (EAE) and Vibriosis.
2. Nasalgen-IP, thirty days or less before lambing. This gives protection against some forms of pneumonia and other respiratory viruses. (Give also to newborn lambs.)

FLUSHING

Flushing is the practice of placing the ewes on an increasing plane of nutrition, that is, in a slight weight-gain situation, to prepare for breeding. It is not as effective if the ewes are fat to begin with. Flushing can be accomplished by supplementing the usual summer diet with grain (and sometimes a better pasture, too). It is most productive when initiated seventeen days prior to turning in the ram, and continued, tapering off gradually, for about thirty days. No advantage is shown by starting it earlier. This system not only gets the ewes in better physical condition for breeding, but it also helps to synchronize them by bringing them into heat at about the same time, preventing long, strung-out lambing sessions.

It is also a factor in twinning, possibly because with this better nourishment the ewes are more likely to drop two ova. The USDA estimates that flushing results in an 18 to 25 percent increase in the number of lambs, and some farmers think it is even more.

You can start with 1/4 pound of grain a day per ewe, and work up to 1/2 or 3/4 pound each in the first week. Continue at that quantity for the seventeen days of flushing. When you turn in the ram, taper off the extra grain gradually.

The ewes will probably come into heat once during that seventeen days of flushing, particularly if you have put the ram in an adjoining pasture, but it is best not to have the ram with them yet, for in the second heat they drop a greater number of eggs, and are more likely to twin if bred during this second cycle.

The ewes should not be pastured on red clover, as it contains estrogen and lowers lambing percentages. White clover may have somewhat the same effect.

Researchers at the University of Illinois have a new "flushing theory" which they say promotes increased ovulations, thus increasing the number of lambs produced. They claim that a daily oral dose of just less than two ounces of mineral oil, given for ten days before mating, will decrease the steroid secretions that normally restrict the number of ovulations. The mineral oil, which in that quantity is said to

Suffolk x Dorset ram, in "courtship" pose (Photo by ram's owner, Jim Gloe)

have no effect on digestion, would be in addition to the normal seventeen-day flushing with supplemental grains and/or better pasture.

I would point out that mineral oil given orally does have its risks. Since it isn't "wet," it doesn't stimulate the swallowing or coughing reflex, and therefore it can be dangerous to give by drench. One can literally pour it down the trachea without so much as a struggle from the animal (until it suffocates).

EWE LAMBS

The exception to the flushing would be the ewe lambs, if you decide to breed them. They will not have reached full size by lambing time, so you would not want them to be bred too early in the breeding season. Don't breed them until the following month. Breeding season is shorter for ewe lambs than for mature ewes, and usually starts in September or October instead of August. Some breeds are slower maturing, like Rambouillet, and some much faster, like Finnsheep, Polypay, and Romanov.

Ewes who breed as lambs are thought to be the most promising, as they show early maturing which is a key to prolific lambing. Ewe lambs should have attained a weight of 85 to 100 pounds by breeding time, as their later growth will be held back a little as compared to unbred lambs. If not well fed, their reproductive lifetime may be shortened, and unless they get a mineral supplement (like TM salt), they will have teeth problems at an early age.

If replacement ewes are chosen for their ability to breed as lambs, the flock will improve in the capacity for ewe lamb breeding, which can be a sales factor to stress when selling breeding stock.

CHOOSE TWINS

Choose your potential replacement ewes from among your earlier-born twin ewes. Turn these twin ewe lambs in with a ram wearing a marking harness, or paint-marked brisket (see Chapter 5). The ones that are marked, and presumably bred, can be kept for your own flock. Sell the rest.

> *Ewes yearly by twinning*
> *Rich masters do make*
> *The lambs from such twinners**
> *For breeders do take.*
>   YOUATT, 1837

A ewe lamb that has twins the first time is more valuable than one who lambs with a single, even though ewes with a future history of twinning may only have a single that first time. Still, they pass on both the inherited ability to breed early and to have twins, and they will produce more lambs during their lifetime.

---

*And, I might add...who are bred as lambs

PUREBREDS AND CROSSBREEDS

When both parents are purebreds of the same breed, the lamb is also a purebred.

When each parent is of a distinct different breed, the lamb is a crossbreed. In crossbreeding you get a lamb that can potentially, but not necessarily, have the good points of both the parents, and is usually faster growing. The value of crossbreeding can be determined in practice by comparing the lamb with the two parent breeds, considering particularly the factors that are of importance in your situation: body conformation; wool; prolificacy; rate of growth; or size.

GRADING UP

The use of a good purebred ram on a flock of very ordinary ewes, and keeping the best of the offspring, is called "grading up." If done for several years, keeping the best of the resulting ewe lambs and disposing of the original ewes, you have probably improved the quality of your flock. The actual improvement will depend partly on the ram chosen, and partly on how carefully you select the ewe lambs that you keep for replacements.

THE RAM

While it has always been known that a whole flock can be upgraded by the introduction of a superior ram, such as by using a purebred ram on common grade ewes, the full implications of this theory have not always been clear. The introduction of purebred genes of a particularly advantageous breed is only part of the answer—the rest is heterosis.

HETEROSIS

Heterosis is the hybrid vigor, the increased hardiness and growth performance that is often found in a crossbreed when it is compared to the average of its purebred parents. This hybrid vigor is the major reason for crossbreeding. The other reason is to breed in such a manner as to allow the strong points of one breed to compensate for the weakness of another.

Individual heterosis comes from crossing two different breeds, and results in an average of 17.8 more pounds of lamb weaned than by purebreeding. Maternal heterosis is the crossing of a purebred ram with a crossbred ewe, and causes an average of 18 percent more lamb produced per ewe, according to statistics from sheep specialists at Ohio State University.

BACKCROSSING

With an unusually good ram, you may want to go one step further, breeding your best ewe lambs back to the same ram, which is called "backcrossing," a form of inbreeding. The lambs resulting from this mating should not be bred back to the same ram.

Inbreeding or close breeding, once referred to as breeding in-and-in, is not without risks, but you can always cull out any lambs that show undesirable traits, such as being undersized, the most common fault of inbreeding.

LINEBREEDING

Linebreeding is a form of inbreeding in which sheep are mated in such a way that their lambs will remain closely related to one highly desirable ancestor. The difference from common inbreeding is that the mated animals should both be related to the one unusual ancestor, but unrelated otherwise to each other. The intent is to maintain a close relationship to that one particular outstanding animal, and propagate its exceptional characteristics, not allowing them to be halved in each following generation.

CULLING

By keeping the best of your ewe lambs and gradually using them to replace older ewes, you should realize more profit.

To know which to cull, you need to keep good records, and this necessitates ear tags. Even if you can recognize each of your sheep by name, you are more inclined to keep clear records with tags than without, and you can also be more efficient about it.

Record: fleece weight each year; wool condition; lambing record; prolapses; rejected lambs; inverted eyelids; milking ability; lamb growth; any foot problems or udder abnormalities; any illnesses and their treatment. With a good history of each animal, you know better what to anticipate.

At culling time, review the records, as well as inspecting teeth, udders, and feet. Cull out ewes with defective udders, broken mouth (teeth missing), limpers who do not respond to regular trimming and foot baths, or those with insufficient milk

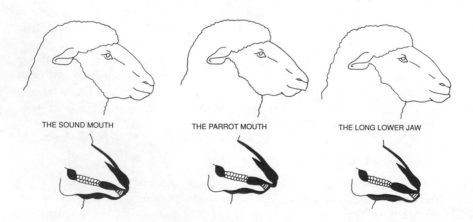

THE SOUND MOUTH   THE PARROT MOUTH   THE LONG LOWER JAW

whose lambs grow slowly. There may be some exceptions to these deficiencies, such as a ewe who regularly has triplets and passes on her prolific traits to her daughters. This one may warrant bottle feeding of her lambs for another season of lambing. Improvements of a flock require rigid culling. Consider all the points listed in Chapter 1, for the purchase of new ewes. Udders, feet, and teeth are always prime areas for inspection. And it is not enough to just have teeth in good condition; the bite itself is important. Dr. Salsbury says, "They can't shear grass if the blades don't match."

Keep in mind especially the ease of lambing—you don't want to end up with a flock of ewes who all require assistance. It is not enough just to select for growth and conformation. If we also select for "survivability," it should show up in a decline of lambing problems.

Shepherd magazine had a good editorial by Guy Flora advocating the "survival of the fittest" type of culling, to propagate the kind of ewe that twins easily, has an excellent mothering instinct, and produces large quantities of nourishing milk from the start. We want to produce lambs that "hit the ground running," with the instinct and strength to stay with their mother, find the milk, suck out the wax plug if necessary, and survive. Be objective and practical, for although the runt you tubed and bottle-fed may be adorable, it is not a viable choice for breeding stock.

TWO LAMB CROPS PER YEAR

The profitable possibility of attaining two lamb crops a year (without use of hormones) is now much closer because of work done by the Animal Science Department of Utah State University. Scientists there devised and tested a method to overcome the common problem of uterine debris that prevents ewes from breeding back early enough to have two lamb crops in twelve months.

Dr. Warren C. Foote explained that infusing the uterus with 200 ml saturated sucrose solution* via the cervix, within four days of lambing, obtained beneficial response. He said that this definitely proved effective in preparing the ewes to breed. Sterile solution and a sterile procedure would be essential to avoid serious complications.

The breeds most noted for out-of-season breeding are the most likely candidates for the practical application of this method. The Polypay breed was developed specifically for the feature of twice-a-year lambing, so it could be at the top of the list of breeds, needing only a little more help to be reliable double producers. Dorset and Finnsheep are good for out-of-season breeding, and those with a moderate capacity for it would be Targhee, Tunis, Panama, and Romanov.

The breed involved in the Utah tests was a Targhee type.

Any program of accelerated lambing will require very early weaning of lambs to prepare the ewe for her next lambing.

---

*Concentrated sucrose solution is a form of sugar and water, obtained by stirring sugar into boiling water, adding as much sugar as will dissolve in it. When cooled, the liquid decanted off the top is concentrated sucrose solution.

CHAPTER 7

Pre-Lambing and Lambing

FEEDS

DO NOT OVERFEED EWES during the early months of pregnancy. A program of increased feeding must be maintained during late gestation to avoid pregnancy disease and other problems. Overfeeding early in pregnancy can cause ewes to gain excessive weight that may later cause difficulty in lambing.

Have adequate feeder space (approximately 20 to 24 inches per ewe) so that all the ewes will have access to the feed at one time; otherwise, timid or older ewes will get crowded out. If possible, they should have free choices of a mineral-salt mix containing selenium. This can make it unnecessary to inject selenium prior to lambing (to protect lambs from white muscle disease). NEVER use a mineral mix intended for cattle because it may be fortified with copper at levels that are toxic to sheep. Some geographical areas require selenium supplementation above the legal limits available in commercial mineral supplements. Check with your local veterinarian or extension agent.

FEEDING IN THE LAST FOUR OR FIVE WEEKS BEFORE LAMBING

By the fourth month of pregnancy, ewes need about four times as much water as they did before pregnancy. And, since 70 percent of the growth of unborn lambs takes place in this last five-to-six-week period, the feed must have adequate calories and nutritional balance to support that growth. During the last month of gestation, the lamb fetuses become so large that they displace much of the space previously occupied by the rumen. This necessitates more high-protein feed and less roughage feed, as the ewes are not able to ingest sufficient roughage, or large quantities of any low-energy feed, to support themselves or the growing lamb(s), which causes them to utilize excessive quantities of stored fat reserves, and can in turn lead to pregnancy toxemia. Poor energy supplementation can also result in hypoglycemia (lowered blood sugar), which mimics the symptoms of pregnancy toxemia. Pregnancy toxemia is

NOT necessarily a "thin ewe" problem. A good grain mix would be 1/3 whole oats, 1/3 shelled corn, and 1/3 wheat (for the selenium content). Barley is a good feed in areas where available. Grain rations can be supplemented to 12 to 15 percent protein content with soybean meal or other protein source. Grain and hay should be given on a regular schedule, to avoid the risk of triggering pregnancy disease or enterotoxemia by erratic eating. Approximately one pound of grain per day (more for larger ewes) is a good rule of thumb.

Poor feeding in the last four weeks (last five to six weeks for twinning ewes) leads to:

- Low birth weight of lambs.
- Low fat reserve in newborn lambs, resulting in more deaths from chilling and exposure.
- Low wool production from those lambs as adults.
- Increased chances of pregnancy toxemia.
- Shortened gestation period, some born slightly premature.
- Ewes slower to come into milk, and less milk.
- Production of "tender" layer (break) in ewe's fleece. This is a weakness that causes the fibers to break in two with the slightest pull, and decreases the wool value.
- *Excessive feeding* can result in excessive growth of the lambs and an overweight condition in the ewe, which can lead to serious lambing problems.

At this time, watch for droopy ewes, ones going off their feed or standing around in a daze. See Chapter 12 for symptoms and treatment of pregnancy toxemia. Both exercise and sunlight are valuable to a ewe that is carrying a lamb, and lack of exercise is one factor in pregnancy disease. If necessary, force exercise by spreading hay for them in various places on clean parts of pasture, once a day, to get them out and walking around.

FEEDING TIME

While regular feeding time is important, several tests suggest that it does make a difference what *time* you feed pregnant ewes. Either late afternoon (shifting even later as lambing approaches), or, about 10:00 a.m., would encourage more daylight lambings. Mid-day feeding gave more night lambing.

THE KETONE TEST

The one way to be sure that your prolific ewes, the ones carrying twins or triplets, are getting enough nutrition (energy) is to check for ketones in the urine. Better to avoid pregnancy toxemia (ketosis) than to be forced to treat it later as an emergency.

Ewes who are not getting enough feed to meet their energy (caloric) requirements will use reserve body fat. When fat cells are converted into energy, waste products called ketones are created. Pregnancy disease (ketosis) results when the ketones are produced faster than they can be excreted and they rise to toxic levels in the bloodstream, which can be easily detected in the urine. A simple test kit for ketones, available at a pharmacy, can be used to identify ewes with caloric deficiencies. Use the ketone test results to separate the ewes that need extra feed, thus avoiding underweight or dead lambs and pregnancy toxemia problems.

VACCINATIONS

Another means of avoiding disease is the appropriate use of vaccines. Those that you may want to use prior to lambing are:

- EAE-Vibrio combination vaccine, which protects against the two most common disease-caused abortions, Enzootic Abortion of Ewes and vibriosis (now called Campylobacter).
- Naselgen-IP (P13 vaccine), which protects against certain types of viral pneumonia and respiratory ills. This vaccine can also be given to newborn lambs quite easily, as it is an intranasal vaccine.
- Covexin-8, which immunizes against all the clostridial diseases, including tetanus and enterotoxemia (Overeating Disease). Follow label directions for the primer dose and booster schedule. This protects both the ewe and the lamb(s) by passing immunity via the colostrum to the newborn lambs during the first fifteen hours of life. Lambs may then be vaccinated, beginning at nine or ten weeks of age, to continue their protection.
- Selenium-E, sub-cu, one week or more prior to lambing, if ewes have not had selenium-enriched feed ration or mineral salt with selenium.
- Ovine Pili Shield, the new newborn-lamb-scour vaccine. One dose, thirty to sixty days prior to lambing, protects lambs through the colostrum against scours caused by *E. coli** bacteria.

SHEARING BEFORE LAMBING

If weather is mild and you do your own shearing so you can be gentle with them, ewes can be sheared up to three or four weeks before lambing. See Chapter 17 for shearing. There are some advantages in having ewes sheared before lambing:

- No dirty, germ-laden wool tags for lambs to suck.
- Clean udder makes it easier for lambs to find teats.
- Fewer germs in contact with the lamb as it emerges at birth.
- Easier to assist at lambing, if necessary.
- Easier to spot an impending prolapse, in time to save ewe (see Prolapse, Chapter 12).
- Easier to predict lambing time by ewe's appearance.
- Ewe less apt to lie on her lamb in pen.
- Shorn ewe requires less space in barn, at feeding racks, and in lambing pen.
- Shorn ewe not as apt to sweat in lamb pen and contract pneumonia.
- Shorn ewe will seek shelter for herself and lamb, in bad weather.

---

* Escherichia coli

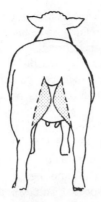

Darkened area is area for crotching.

CROTCHING (TAGGING) BEFORE LAMBING

Actually, the first five advantages of pre-lambing shearing are gained also by *crotching* (sometimes called *crutching*), which is trimming wool from the crotch and udder and a few inches forward of the udder on the stomach. Only about four or five ounces of low-value wool are removed, and this can be washed for spinning or sold with the fleece.

FACING

(called "wigging" in some countries)

Another practice of value before lambing, or before the ewe is turned out of the lambing pen, is "facing" (trimming the wool off the ewe's face). It is often done while crotching. It has several purposes. In "closed-faced" sheep (sheep with heavy wool about the eyes and cheeks), it avoids "wool blindness," enables the ewe to locate and watch her lamb more easily, and helps prevent the accumulation of hay chaff and burrs in the wool while eating hay at the bunk. With "open-faced" sheep, it still makes the ewe more likely to seek shelter with her lamb in wind and rain, even if she has not been sheared.

LAMBING PENS ("JUGS")

Have a 4-x-6-foot (or 6-x-6 for large-breed sheep) lambing pen, "claiming pen," or "jug" ready for the newborn lamb and its mother, with clean bedding, a small hay feeder, and a container of water that cannot be spilled and is tall enough that a small lamb cannot fall into it and drown. A plastic 5-gallon pail is ideal but don't let the bedding build up around it so that a lamb can walk into the bucket and drown! As a general rule you will need approximately one jug for every ten ewes in the flock. If you own a small flock, you should be prepared to have a minimum of three jugs.

Give thought to the barn environment. A healthy barn must not be warm, but should be clean, dry, and free of drafts. Warm or drafty barns can cause pneumonia in the young lambs. A warm, damp barn is extremely conducive to bacterial growth.

A closed barn without proper ventilation allows ammonia from fecal decay and urine to build up. This excessive ammonia can irritate the lining of the lungs and trachea, predisposing an animal to severe pneumonia and respiratory disease. Barns should be cleaned out and well dried in preparation for lambing season.

Ewes lambing for the first time, especially yearlings, can be nervous or confused because they do not have the experience of previous lambings, or the fully developed maternal instincts. They should be "jugged" (penned) with their lambs for at least three days until they become accustomed to the nursing lamb(s).

If the lamb cries a lot, that is one indication that it is hungry, but not always. Lambs will sometimes starve to death in the jug without a sound! Check milk DAILY for the first three days—that she does have milk and the lamb is getting some. Some ewes may come to milk only to dry up after a day or two, so never assume that a ewe will continue to milk after the first day.

If a young ewe does not have sufficient milk for the lamb, supplement it with a couple of 2-ounce bottle feedings for the first two days, preferably with milk taken from another ewe or with newborn milk formula (see Chapter 10). Insufficient milk letdown can sometimes be resolved by injections of Oxytocin available from your veterinarian.

The ewe's milk should increase if she is well fed. If it still is not sufficient for the lamb, supplement it with a couple of 4-ounce feedings of lamb milk-replacer during the first week, then increase to about 8-ounce feedings at two weeks old. Poorly fed old ewes also may have scant milk supply.

The jug is primarily for use after the lamb is born. Ewes prefer a larger area for the actual lambing, where they can walk around freely before labor. Because of the trend toward larger sheep, recommendations for jug size have been increasing upwards. The larger pen (6 X 6 feet) is best if you want to have the ewe confined where facilities are better for helping in a difficult birth. If weather is bad, you may want to have the ewe in the pen as labor starts, where there is good light to watch her progress.

The jug allows the ewe and lamb to become well acquainted without distraction, keeps the lamb from getting separated from its mother (especially in the case of twins or triplets where the ewe cannot count beyond "one"), and protects the lamb from being trampled by other sheep or becoming wet and chilled. Ordinarily, they are penned together for three days so that they can be easily observed and treated, should complications arise. Do NOT allow dogs or strangers to approach the jug area, especially with nervous ewes. A frightened or nervous ewe can quickly turn a serene protective jug into a "lamb blender" with fatal results.

If the ewe lambs outside, it is not difficult to get her to the jugs nearby. Carry the lamb slowly, close to the ground so she can see it and follow. Since lambs do not "fly," the ewe will instinctively look for the lamb on the ground. If the lamb is raised more than a foot or so off the ground, the ewe may "lose" it and run back to where she dropped it. When this happens, you will need to go back and begin again. If the lamb calls out to her along the way, she will normally follow readily. There are commercial "lamb cradles" and "lamb packers" available that allow you to carry the newborn lamb inches off the ground as if it were a suitcase. The "Norseman-

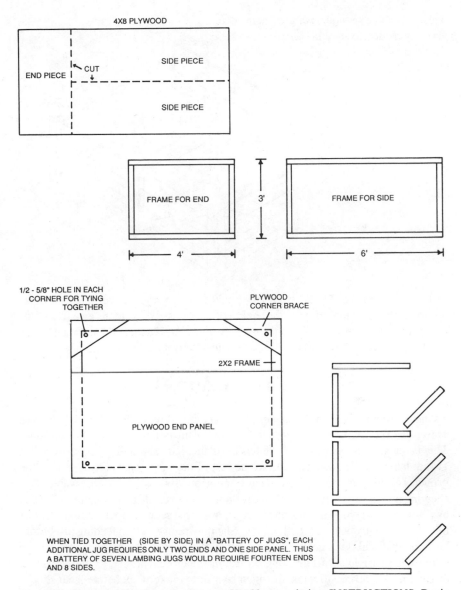

WHEN TIED TOGETHER (SIDE BY SIDE) IN A "BATTERY OF JUGS", EACH
ADDITIONAL JUG REQUIRES ONLY TWO ENDS AND ONE SIDE PANEL. THUS
A BATTERY OF SEVEN LAMBING JUGS WOULD REQUIRE FOURTEEN ENDS
AND 8 SIDES.

Norseman Sheep Co. Lambing Jug Pattern. Used by permission. INSTRUCTIONS: Begin
with a 4 x 8 sheet of 1/2-inch waterproof sheathing plywood. Cut a 2-foot piece from the end
of the sheet, and then split the remaining piece down the center. This results in one piece that
is 2 feet by 4 feet, and two pieces that are 2 feet by 6 feet. These are nailed to 2 x 2 frames
that are 3 feet by 4 feet and 3 feet by 6 feet respectively. A single sheet of plywood will make
one end and two side pieces, or since one uses more ends than sides, four ends can be cut from
a single sheet.

When tied together (side by side) in a "battery of jugs," each additional jug only requires
two ends and one side panel. Thus a battery of seven lambing jugs would require fourteen
ends and eight sides.

Folding lambing pens (jugs) in place in the barn.
Ewe hay racks are shown in each pen.

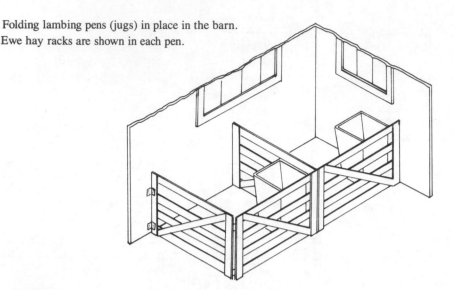

pattern" lamb packer fits easily into the pocket and is very convenient to use. It has
the advantage of being easier on the back, with less distraction to the ewe from your
humped-over appearance. From the ewe's viewpoint, it appears that the lamb has
suddenly begun to follow you and she will instinctively follow it.

LAMBING SYMPTOMS

As the time approaches for actual lambing, the lamb(s) will "drop," giving the ewe
a sway-backed, sunken appearance in front of the hip bones, and a restless attitude.
This is much more noticeable if she has been sheared. She will often pick out her
spot to lamb and lie down apart from the rest of the sheep, sometimes pawing the
ground before lying down. Too much lying around without any observable cud-
chewing may be an early sign of the droopiness of toxemia (see Chapter 12). The
ewe will normally have made a bag by now, but some seem to hold out until the last
minute or actually bag out after lambing. She may refuse a grain feeding just before
lambing, but will often gobble up an apple cut in pieces. Our ewes who are huge
with twins or triplets start grunting several days before lambing, as they lie down
or get up. The vulva will relax and often be a little pinker than before—but should
not be protruding and red, which could be the beginning of a prolapse (see Chapter
12).

LAMBING

Keep your fingernails trimmed close, in case you have to assist in delivery, and have
the following pre-lambing supplies on hand. Many are obtained by mail. Order them
ahead of time so you will have them when needed. Here is a list, not necessarily in
the exact order of importance or in the sequence in which they may be needed:

1. Roll of paper towels in lambing pen.
2. Old terry towels for drying off lambs. Store in plastic bags to keep clean and dry.
3. Electric hair dryer for warming and drying lambs in cold weather.
4. Strong (7 percent) tincture of iodine (not "tamed" iodine) in a small wide-mouth jar, for treating umbilical cord.
5. Small sharp scissors for trimming umbilical cord.
6. Hand sheep shears for crotching. Can also be used for annual shearing.
7. Antiseptic and lubricating ointment for your hands, if you have to assist in delivery. Cooper's Dairy Ointment or Septi-Lube are good.
8. Clorox-brand laundry bleach for disinfecting hands and equipment. NOTE: Other commercial brands of chlorine bleach contain ingredients that can be quite irritating to the skin.
9. Antibiotic uterine boluses in case of retained placenta.
10. Sterile syringes and disposable needles, 18-gauge.
11. Penicillin ("Pen-Strep," Combiotic), Liquamycin, or LA 200.
12. Lambing "snare" to help pull lamb in difficult delivery.
13. Heavy cotton or nylon line for lambing loops (snares). Dip in antiseptic or "Clorox" solution before using.
14. Livestock molasses (grocery store kind is too expensive; get this at feed store).
15. Propylene glycol for treatment of pregnancy toxemia.
16. Baby bottle with slightly enlarged nipple hole for newborn lamb.
17. Heat lamp with ceramic-base holder, and heavy-duty extension cords if pens not adequately wired.
18. Frozen colostrum (thaw at room temperature if needed) or newborn lamb formula (Chapter 10). Make up if needed.
19. Mineral oil in case of constipated lamb.
20. Pepto Bismol for simple diarrhea caused by overfeeding.
21. Elastrator pliers with rubber rings, for both castration and docking of tails. Dock tails at two to three days old, depending on vigor of lambs.
22. Calcium gluconate for treatment of milk fever.
23. Lamb ear tags and applicator.
24. Record book and hanging scale to weigh lambs. (Old baby scales work well.)
25. Ear syringe.
26. Alcohol and cotton.
27. CLEAN plastic bucket.
28. Prolapse retainer and prolapse harness (see Sources).
29. Sulfa preparation for lamb scours (diarrhea). Those made for baby pigs work great.
30. Bucket of warm water for ewe to drink.
31. Rectal thermometer.
32. 5 percent or 50 percent glucose in saline solution.
33. Lambing pens with feed and water equipment.
34. Squeeze bottle of liquid soap.
35. Powdered lamb milk-replacer (calf milk-replacer will not substitute).
36. Colostrum powder or Colostryx (new antibody supplement).

START OF ACTUAL LAMBING

Labor is beginning when the ewe lies down with nose pointed up, then strains and grunts. Give her plenty of time to lamb by herself before trying to assist, unless the lamb is showing and she is making little progress. You can pull the lamb, timing your pulls with her straining.

Most veterinarians recommend that you allow one-half to one hour after the water bag comes out, or one and one-half to two hours of labor, before examining the ewe or assisting, but you'll have to judge from her appearance as to whether the ewe is becoming so tired that she needs assistance.

In the great majority of cases, she will give birth normally and easily. If you have any lambing problems, see Chapter 8 for instructions on how to deliver lambs.

At birth. If you are there when the lamb is born, wipe the mucus off the lamb's nose, then place it at the ewe's head quickly so she can identify it as her own and clean it off. (Now is the time to "graft" on an orphan or triplet that needs a foster mother. See instruction in Chapter 10).

If the lamb has difficulty breathing or excess mucus in the throat and lungs, grasp it firmly by the hind legs and swing it aggressively in an arc several times in order that centrifugal force will expel the mucus. Make sure that you have a good grip on the lamb to avoid throwing it out of the barn and make sure that its head will clear the ground and all obstacles!

If the navel cord is over 2 inches long, snip it off with scissors and submerse it in the 7 percent tincture of iodine solution. Have the iodine in a small wide-mouth container. Hold the lamb so that the navel cord hangs into the container. Press the container against the lamb's belly, then turn the lamb up so that the entire cord and the area surrounding it are covered. Iodine should be applied as soon as possible after birth, because many bacteria can enter via the navel. The iodine penetrates the cord, disinfecting it, and assists in drying. Avoid spilling the iodine on the lamb or applying it to an excessive area, because it has a strong odor that may mask the lamb's natural odor and cause the ewe to reject it. As an extra precaution against infection, you can treat the cord with iodine again in twelve hours.

If the cord is not cut to the proper length, some ewes try to nibble too much of the navel and can injure the lamb. One year we had an excited ewe who chewed the tail off her newly born lamb, nibbling it as if it were an umbilical cord. In over thirty years, this had never happened. We put a band on the tail, above where she had chewed it, and dunked it in iodine, then went to get warm molasses water for her to drink. When we got back, she had just had a second lamb, and we couldn't believe it. She chomped the tail off that one, too! Nice that she was in the lambing pen with a good light, so we could see what had happened, and could take care of those poor tails. She licked off the lambs, and was a wonderful mother. If this had happened with more than one ewe, we would have suspected a nutritional deficiency, for that is reported to be one sign of it.

In cold weather, you must guard the newborn lamb against hypothermia. Once dry they can withstand quite low temperatures, but due to a large ratio of skin area to body weight, wet lambs can chill quite quickly. A hypothermic lamb will appear stiff, will be unable to rise, and its tongue and mouth will feel cold to the touch. You must warm it immediately with an outside heat source, because it has lost its ability to control its temperature. Wrapping it in a towel or blanket will NOT suffice.

The best method of warming a "frozen" lamb is to submerse it up to its neck in water that is quite warm to the touch. (See "Lamb Problems," Chapter 11.) Most lambs will revive in just a few minutes. When the mouth begins to feel warm to the touch, and the lamb begins to struggle, dry it well and place it in a warm environment until totally recovered. Feed it one to two ounces of warm colostrum or milk-replacer as soon as it can take it. If you are experienced, force-feeding with a stomach tube after removal from the water and drying will speed up the recovery.

If the ewe is too exhausted by a difficult labor to dry off the lamb, do it yourself with paper towels, so that the lamb does not get cold from being wet too long. Do not remove the lamb from her sight, as this can disrupt the mothering-ownership

The author's Lambie Pie and one-day-old lamb.

pattern. Allow her to lick off the amniotic fluid, but even if she is not able to, put the lamb near her nose to encourage her to identify with it. Overuse of a heat lamp to dry the lamb can result in a "chill" when removed and predispose it to pneumonia later.

The use of plastic "lamb coats" in cold weather can be beneficial because they retain a great deal of body heat. A newborn lamb will appear "wrinkly" because there is very little body fat under the skin. It takes approximately three to five days to build up that fat layer under the skin, which acts as natural insulation. When a plastic lamb coat is used to help the lamb retain body heat, the energy that would be used to keep it warm is converted to body fat. This can be especially beneficial to twins and triplets on marginal milk intake.

If the lamb should be born dead, you can rub a young orphan lamb all over with the birth fluid and give it to the ewe to mother. (See Chapter 10.)

Unplug the ewe. Strip the teats of the ewe to unplug her, as the lamb may not suck strongly enough to remove the little waxy plug.

Check eyelids. Check the lamb's eyelids to see if they appear to be turned in so that the eyelashes would irritate the eye. This can cause serious trouble and blindness if it is not corrected, and the sooner it is noticed the easier the remedy. (See Entropion, Chapter 11.)

NURSING

When the ewe stands up, she will nudge the lamb toward her udder with her nose, if it is strong enough to get on its feet. The lamb is born with the instinct to look for her teats, and also is drawn by the smell of the waxy secretion of the mammary pouch gland in her groin. If the udder or teats are dirty with mud or manure, a swab with a weak chlorine (Clorox) solution before the lamb nurses will clean things up and help prevent intestinal infection in the lamb.

Let the lamb nurse by itself if it will, but do not let more than a half-hour to one hour pass without it nursing, as the colostrum (ewe's first milk after lambing) provides not only warmth and energy, but also antibodies to the common disease organisms in its environment.

Occasionally the ewe will not allow the lamb to nurse because she is nervous, has a tender or sensitive udder, or is rejecting the lamb. If the udder appears sensitive, it is often because it is tightly inflated with milk. Restrain the ewe and allow the lamb to nurse. You can then milk out the excess colostrum (save it if possible) to remove the pressure on the udder. Nervous ewes may require restraint for the first few feedings, until they get the hang of being a mother. A rejected lamb is well on its way to being an orphan ("bummer") lamb. (See "Orphan Lambs," Chapter 10.)

You will have added greatly to the colostral protection of the newborn lamb if you have previously vaccinated the ewe (twice) with Covexin-8 to protect against tetanus, enterotoxemia, and the other common clostridial diseases (see "Vaccinations," this chapter). These antibodies are absorbed by the mammary gland from the ewe's bloodstream and are incorporated into the colostrum so that they protect the newborn lamb until it starts to manufacture its own antibodies. The small intestine of the newborn lamb possesses the very temporary ability to absorb these large molecular antibodies from the colostrum. This ability to absorb colostrum decreases by the hour until it is almost nonexistent by sixteen to eighteen hours of life. Colostrum is also high in vitamins and protein, and is a mild laxative to pass the fetal dung (meconium, the black tarry substance that is passed shortly after the lamb nurses).

The longer a lamb has to survive without colostrum, the fewer antibodies it has the opportunity to absorb, and the less its chance of survival if it develops problems. A weak lamb or one of light birth weight can be lost because of a delay in nursing.

This is not the same as loss due to starvation, or from receiving no milk at all, as a strong lamb can sometimes survive for a day or more without ever getting any milk, but getting weaker all the time. Many lamb deaths that are attributed to disease are actually due to starvation, and lambs will often die having not uttered a sound or indicated that they were starving. Always make sure that the lambs are actually nursing, and always recheck the ewe to make sure she is continuing to give milk for the first few days.

We don't usually wait around for the lamb to nurse, but just roll the ewe on her side and press the lamb's nose against her teat. If it does not readily suck by itself when it feels the warm udder, we push the teat into its mouth, from the side. It usually cooperates, getting the urge when it feels the warmth in its mouth. After this first feeding, we have some assurance that it will have the strength to look for the next one, but we keep watch to make sure that it does nurse from time to time.

On occasion you may encounter the "lazy" lamb, that for no apparent reason does not want to nurse the ewe, but will take a bottle with enthusiasm. These lambs can be maddeningly frustrating and can tax both your patience and your nerves as to how long you are willing to stand the suspense to see if it will begin nursing the ewe or not. We call these lambs "volunteer bummers." If you have a heavy lambing schedule, see Chapter 10.

MOLASSES AND FEED FOR MAMA

Ewes are often thirsty after giving birth. We offer the ewe a large bucket of warm water (not hot) containing half a cup of stock molasses. It is important to have it warmed, as the ewe may be reluctant to drink very cold water which can result in lowered milk production. Offer good hay, but no grain the first day, as it could promote more milk than a tiny lamb could use. If she has twins or triplets, however, and seems short of milk, grain feeding should start that first day.

If she has too much milk, her udder is too full and the teats are enlarged from it, milk out a bit of this colostrum and freeze it in small containers for emergency use. Ice cube containers or small "zip-lock" bags are very good, as they allow the thawing of small quantities as needed. Solidly frozen colostrum will keep for a year or more, if well wrapped. When saving and freezing colostrum, it is even better to have a combination of colostrum milked from several ewes, for they do not all produce the same broad spectrum of disease-fighting antibodies. Cow or goat colostrum can be stored and used in emergencies.

Thaw frozen colostrum at room temperature or in lukewarm water. Never use hot water or a microwave oven to thaw colostrum because it can denature and destroy the antibodies, rendering the colostrum worthless.

TWINS

Twins require vigilance to assure that both lambs are claimed by the ewe, and that each is getting its share of colostrum. If the ewe does not have plenty of milk for them, increase grain gradually. Unless she shows some reluctance about the molasses, continue offering it in lukewarm water during the time she is jugged with the lambs.

If twins cry a lot, they are probably not getting enough milk. Notice if only one of them is crying, or both, and assist the hungry one by holding it to its mother. If she is short on milk for both, give a supplemental bottle. When a ewe does not have enough milk for multiple lambs, you can still leave them all nursing her, but supplement one of them or all of them with a couple of bottle feedings a day. Give two-ounce feedings the first couple of days, and increase to four to five ounces by the third and fourth day, gradually increasing as they grow, if her milk is still not adequate. See Chapter 10 for the newborn lamb milk formula to feed for the first two days, then gradually change to lamb milk-replacer (not calf milk-replacer). A brightly colored small nylon dog collar or collar made of yarn is a convenient way to "flag" any lambs that need special observation, as they really stand out among the mob.

Tarhgee and triplets. (Mt. Haggin Livestock, Inc., Anaconda, MT)

LAMB DROPPINGS

One advantage of penning lambs with their mothers is that you can keep an eye on how well they are eating, and on how well it is coming out the other end—the condition of the droppings is important.

First to come out is the fetal meconium, a gob of black tarry matter, which is passed a few hours after the lamb is born. This is the remnant of the amniotic fluid that is swallowed and digested by the fetus while developing in the "water bag" before birth. The next droppings are bright yellow, but the same consistency. They remain yellow for at least a week, then gradually get darker until they are a normal brown small bunch of pellets sticking together in clumps like small pinecones. Later, they are little brown marbles. If they become loose and runny, this is called "scours." See Chapter 11 for treatment.

EAR TAGS

If you have more than two or three ewes, which should produce two to six lambs, you can identify the lambs best by ear tags. This makes it possible to keep records of lamb parentage, date of birth, and growth, and easier to decide what to keep for your flock and what to sell. With identification tags on your ewes also, you can be certain which lambs are hers, even after they are weaned.

Livestock supply catalogs sell a variety of tags. Some are metal with almost any combination of numbers and letters (your name if you wish), and some are plastic in a variety of colors, also with numbers and letters of your choice. The different colors can be used to identify sex, whether twins or singles, the month born, etc. Some are a self-clinching type, while others need a hole punched for the tag. These should be applied while the lamb is still penned with its ewe. Never use large heavy "cow" tags on adult sheep. Similarly, tags intended for mature sheep are often too heavy for a lamb's ear to support. If using the small metal lamb tag, it should only be inserted onto the ear approximately half the length of the tag in order to leave growing room for the maturing ear.

CHAPTER 8

Abnormal Lambing Positions and How To Help

USUALLY THE EWE will give birth unassisted, but you should be prepared for the abnormal delivery. During lambing season keep your fingernails cut short, in case emergency requires you to reach in to pull a lamb. You also will need the following supplies readily available if it is necessary to help the ewe when lambing.

SUPPLIES TO HAVE ON HAND

(see also list in Chapter 7)

1. Good light in the delivery area.
2. A lambing snare, or several pieces of strong cord, with a noose on the end of each one.
3. Antiseptic lubricant or mineral oil.
4. Bucket of clean soapy water to wash your hands and arms, and external parts of the ewe.
5. Penicillin to give after assisting, if in doubt that sterile procedure was used.
6. Roll of paper towels.
7. Iodine in small wide-mouth bottle.
8. Antibiotic uterine boluses.

HELPING THE EWE

It is always a quandary to know when to help and when not to help. As a general rule, you can allow a half-hour to an hour after the water bag breaks, or one and one-half to two hours of labor, before you try to determine the position of the lamb. You want to give her time to expel it herself if she can, but not wait until she has stopped trying.

The size of the pelvic opening is usually large enough for the lamb's body to come out if it is in the normal position, with the front legs and the head coming first. If it is not in this position, delivery is seldom possible without some repositioning of the lamb, or assistance.

After you have washed your hands and arms, and washed off the ewe, you can lubricate one hand and slip it in gently, to try to find out the position of the lamb.

Identify the lamb's legs, and position. First, make sure that the legs you feel belong to the same lamb. In twin births, frequently one or both of the lambs come backwards, and it's easy to get their legs mixed up.

The front legs, above the knees, have a muscular development. The hind legs have a prominent tendon. The front knee bends the same way as the foot (pastern) joint, with the knuckle pointing forward. The back knee joint bends the opposite way from the back foot, and has a sharper knuckle, pointing backwards. If you have a small lamb, catch it and feel the difference between its front and back legs.

When repositioning a lamb to change an abnormal position, avoid breaking the naval cord, as the lamb will attempt to breathe as soon as the cord is broken.

When helping, time your pulling to coordinate with the ewe's labor contractions. If she is tired and has stopped trying, she will usually start again when you start pulling on the lamb.

After difficult lambing. Place the lamb at the ewe's nose. She may be exhausted and otherwise might not clean the lamb, an act which reinforces the mothering instinct. If her nose feels that warm, wet lamb, she will usually try to lick it off. You may have to help her get it dry. Then, put iodine on its naval, and assist it to nurse, even if the ewe cannot stand up yet.

When to call the vet. If a ewe is obviously in distress, has labored over an hour with no progress, and you cannot get the lamb into proper position for delivery, call the vet. When you are paying for a vet, be sure you learn all you can. They don't ordinarily explain things unless you ask questions, and show an interest.

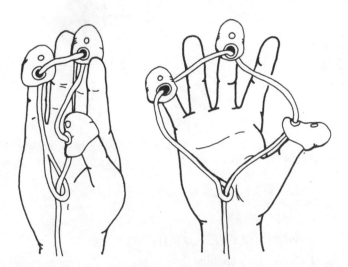

These soft "sensory fingertips" are placed on your own lambing-loop, and are a great help in positioning loop over the lamb's head, for an assisted delivery. To order, see Sources. They can also be made from sturdy industrial-grade rubber gloves.

If a lamb is dead in a ewe, and so large it can't be pulled out, a veterinarian may have to dismember the lamb to remove it.

POSSIBLE LAMB POSITIONS

1. Normal, front feet and head coming out.
2. Large head or shoulders (tight delivery).
3. Front half of lamb out, hips locked.
4. Head and one leg, with one leg turned back.
5. Head, with both legs turned back.
6. Both legs, with head turned back.
7. Hind feet coming first.
8. Breech.
9. Lamb lying crossways.
10. All four legs presented at once.
11. Twins, mixed up, presented at once.
12. Twins, one coming backward, one forward.

1. NORMAL BIRTH

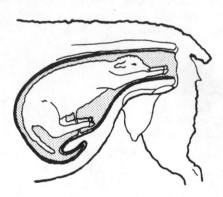

Nose, and both front feet, are presented. The lamb's back is toward the ewe's back. It should start to come out a half-hour to an hour after the ewe has passed the water bag.

She should need no help unless the lamb is large, or has large head or shoulder (see position #2).

2. LARGE HEAD OR SHOULDERS, TIGHT DELIVERY

Ewe may have difficulty lambing, even with the lamb in normal position, if the lamb is extra large, or the ewe has a small pelvic opening.

Sometimes the shoulders are large, and stopped by the pelvic opening. Use a gentle outward and downward pulling. Pull to the left or the right, so shoulders go through at more of an angle, and more easily.

Occasionally the head is large, or swollen if the ewe has been in labor quite a while. Assist by pushing the skin of the vulva back over the head. When the lamb is half-way out (past the rib cage), the mother usually can expel it by herself, unless she already is exhausted.

When the head is extra large, draw out one leg a little more than the other, while working the ewe's skin back past the top of the lamb's head. Once the head is through, you can extend the other leg completely, and pull out lamb by both legs and neck. If both of the legs are pulled out equally, the thickest part of the legs comes right beside the head, making delivery more difficult (on ewe and you).

Pulling gently from side to side assists birth more than only outward and downward movement as in normal delivery.

Use mineral oil, or antiseptic lubricant with difficult, large lamb. Use loop over lamb's head so that the top of the noose is behind the ears, and the bottom of the snare is in the lamb's mouth. Gentle pulling on the head as well as the legs, is better than pulling on legs only.

3. FRONT HALF OF LAMB OUT, HIPS STUCK

This is only a difficult position for the ewe, who may be weary from labor and need help. While pulling gently on the lamb, swing it a bit from side to side, and if this doesn't make it slip out easily, give it about a quarter turn, while pulling. A large lamb in a small ewe will often need this kind of assistance.

4. HEAD AND ONE LEG COMING OUT

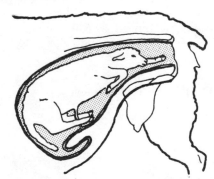

Veterinarians are not in agreement on the procedure to follow when the head and one leg are presented, with one leg still turned back. Some reason that since some ewes can lamb unassisted with the lamb in this position, it should be safe enough to pull gently on the one leg and head, to deliver the lamb, if the ewe is having difficulty. Others say it's risky, and that the folded-back leg may kick at the ewe, causing internal damage.

To change this to a normal birth position, attach a snare-cord to the one leg that is coming out, and also one onto the head. Then push them back enough to enable you to bring the retained leg forward, so you can pull the lamb out in normal position. The cord on the head is important, for the head may drop out of the pelvic girdle, making it difficult to get it started back again.

If the right leg is presented, the ewe should be lying on her right side, so the turned-back leg is uppermost. This would make it easier either to get that backward leg into the right position, or even to help the ewe to lamb even though the leg is not in the normal position.

5. HEAD, WITH BOTH LEGS TURNED BACK

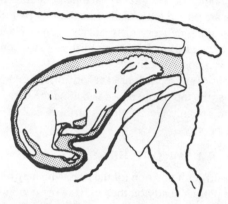

Attach noose onto head (behind ears, and inside mouth). Try to bring one leg down into position, then the other, without pushing the head back any further than necessary. Attach noose of cord onto each leg as you get it out, then pull lamb.

If your hand cannot pass the head to reach the legs, place the ewe with her hind end elevated, which gives you more space. With snare over the lamb's head, push it back until you are able to reach past it and bring the front legs forward, one at a time. Put ewe back in normal reclining position, start head and legs through pelvic arch, and pull gently downward.

6. BOTH LEGS, WITH HEAD TURNED BACK

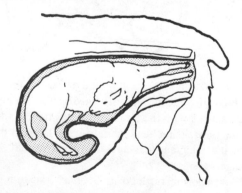

Head may be turned back to one side along the lamb's body, or down between its front legs.

If front legs are showing, slip a noose of heavy cord over each front leg, then push lamb back until you can insert lubricated hand and feel for the head position, then bring head forward into its normal position. With noose on legs, you won't lose them. Pulling gently on legs in downward direction, guide the head so that it will pass through the opening of the pelvic cavity at the same time as the feet emerge on the outside.

If the head does not come out easily, it is either a large head, or the lamb may be turned on its back (with its back down toward the ewe's stomach). With cords still attached to legs, you may have to push it back again, and gently turn it a half turn, so that its legs are pointed down in normal position, for it will come out easier that way.

If you have a hard time getting a grip on that slippery head to bring it into position, try to get a cord-noose over its lower jaw. Insert your hand with the noose over your fingers, then slip it off onto the chin. Be sure it does not clamp down on any part of the inside of the ewe, and tear her tissues. By pulling on the noose that is over the chin, you can more easily guide the head into position.

7. HIND FEET COMING OUT FIRST

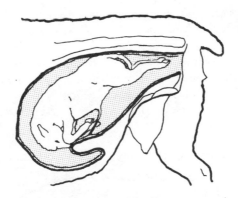

Pull gently, as the lamb often gets stuck when half-way out. When this happens, swing the lamb from side to side while pulling, until ribs are out, then pull out quickly. Wipe off its nose at once so the lamb can breathe. Delay at this point can allow the lamb to suffocate in the mucus that covers the nose. Sometimes it's easier on the lamb if it is twisted one-half turn, so its back is toward the ewe's stomach, or even rotating it a quarter turn while pulling it out. Finish pulling it out quickly because the umbilical cord is pinched once the lamb is half-out, and if lamb tries to breathe, it will draw in mucus.

8. BREECH

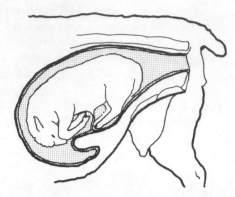

In the breech birth, the lamb is presented backwards, with its tail toward the pelvic opening, and the hind legs pointed away from the pelvic opening. It is generally easier to get it into position for coming out with its back feet coming out first, but once you get it started out, speed of delivery is important. The lamb will try to breathe as soon as the navel cord is pinched or broken, so it can suffocate in mucus if things take too long. Wipe off its nose as soon as it pops out.

To deliver, change breech position by positioning the ewe with her hind end somewhat elevated, so that the lamb inside her can be pushed forward in the womb. This will make barely enough space to reach in and slip your hand under the lamb's rear. Take the hind legs, one at a time, flex them, and bring each foot around into the birth canal.

When the legs are protruding, you can pull gently until the rear end appears; then grip both the legs and the hind quarters if possible, and pull downward, not straight out.

If the ewe is obviously too exhausted to labor any more, try to determine if there is another lamb still inside her. If not, go ahead and give her a penicillin shot, or insert antibiotic uterine bolus, to prevent infection.

However, if you are unable to bring out the legs, and the ewe is making no progress, call the vet.

9. LAMB LYING CROSSWAYS

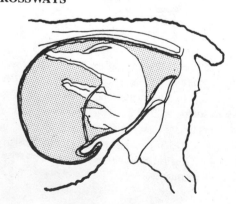

It sometimes happens that the lamb is lying across the pelvic opening, and only the back will be felt. If you push the lamb back a little, you can feel which direction is which. It can usually be pulled out easier hind feet first, especially if these are closer to the opening. If you do push it around to deliver in normal position, the head will have to be pulled around. If it is also upside down, it will need to be turned a half turn to come out easily.

10. ALL FOUR LEGS PRESENTED AT ONCE

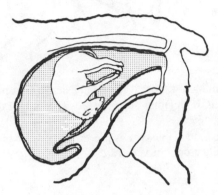

If the hind legs are as convenient as the front, choose the hind legs and you won't have to reposition the head. If you choose the front legs, head also must be maneuvered into correct birth position along with the legs. Attach cords to the legs before pushing back to position the head.

11. TWINS COMING OUT TOGETHER

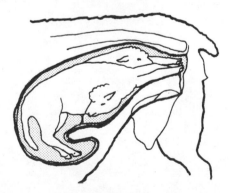

When you have too many feet in the birth canal, try to sort them out, tying strings on the two front legs of the same lamb and tracing the legs back to the body to make sure it is the same lamb, then position the head before pulling. Push the second lamb back a little to give room for delivery of the first one.

12. TWINS, ONE COMING OUT BACKWARD

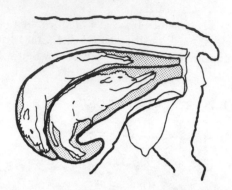

With twins coming together, it is often easier to first pull out the one that is reversed. More often, both lambs are reversed, so you pull the lamb that is closer to the opening.

Sometimes, the head of one twin is presented between the forelegs of the other twin, a confusing situation, but very rare.

CHAPTER 9

Care of Baby Lambs

TO REDUCE REPETITION, we will assume that you have read Chapter 7, "Pre-Lambing and Lambing," and that you have done the urgent things, such as saturating the umbilical cord with iodine, making sure the lamb has nursed, and that it is dry and appears healthy. If the newborn is weak, or has inverted eyelids or other obvious problems, see Chapter 11, "Lamb Problems."

VACCINES

Even though the ewe has had her Naselgen, there is only a limited immunity passed on to the lamb, in this regard. To protect against certain forms of pneumonia to which the newborn lamb is quite susceptible, it should have its own vaccination (intranasal) with Naselgen.

The ewe's primer and booster shots of Covexin-8 will protect her, and will pass on this protection to the lambs from birth until the age of about nine or ten weeks. Since they will still need immunity from tetanus, enterotoxemia, and other clostridial diseases, don't forget to give each lamb its own shot of Covexin-8 by the age of ten weeks.

DOCKING

Tails should be docked (removed) before the lambs are turned out of the lambing pen (jug). This is much easier on the lamb when it is two or three days old and the tail is still small. Sheep of most breeds are born with long tails, and these can accumulate large amounts of manure on the wool, attracting flies and then maggots (fly strike), and can serve as a general source of filth, interfering with breeding, lambing, and shearing.

There are many ways to remove tails, and some are better than others. Docking can be done by cutting with a dull knife (a sharp knife causes more bleeding), a knife and hammer over a wooden block, a hot electric chisel or clamp (this cauterizes the wound to lessen bleeding), a Burdizzo emasculator and knife (which crushes the ends of the blood vessels) or the Elastrator, which applies a small strong rubber ring to cut off the circulation, causing the tail to drop off in a couple of weeks. I have always favored the Elastrator because it minimizes shock and eliminates bleeding problems.

My veterinarian friend prefers the Burdizzo method. Each has advantages, depending on the circumstances of use and the person concerned.

The Elastrator prevents bleeding, is very economical in terms of supplies and equipment needed, and is the easiest method for the beginning shepherd to learn and use. The main disadvantage could be the risk of tetanus, if you have not vaccinated. (This is a risk with all methods, however, without vaccination.)

The Burdizzo method is quick. If accompanied by a mattress-type suture of the skin, it is almost bloodless and the wound would be almost healed about the time the tails had dropped off had you used the Elastrator. The disadvantage of the Burdizzo is the need for more expensive equipment, a suture procedure, and greater operator skill.

What length is best for the tail? It is stylish among purebred producers to cut tails off at the base, leaving practically no tail stub. (My veterinarian says that they would remove them just behind the ears if they could!) However stylish as "no-tail" docking might be in the show ring, the damage to the tissues surrounding the anus definitely predisposes ewes to rectal prolapses. A farm producer should leave the tail longer, cutting or applying the band at the third joint, which is about 1 inch or 1 1/2 inches from the body. As you lift the tail, you will notice two flaps of skin that attach from the underside of the tail to the area on each side of the rectum. The band or cut should be placed just at or slightly past where the skin attaches to the tail (on the tail, not the skin). This leaves enough tail to serve as a cover, and prevents damage to the muscle structure that could weaken the area and add to the risk of prolapse later on.

Whatever procedure you use, be clean. The Elastrator rubber rings should be stored in a small wide-mouth jar of alcohol, disinfectant, or mild Clorox solution to keep them sterile and to disinfect your fingers when you reach for one. Dip the Elastrator pliers, Burdizzo, and/or knife in it, too. With the Elastrator, the tail falls off in one to two weeks, but after three days it can be cut off, on the body side close to the band, and the stump dunked in 7 percent iodine.

Some people object to the Elastrator docking, believing the lamb is more prone to tetanus. This could be true, but tetanus can result from any method of docking. No farm is free from tetanus and those who fail to prevent it will sooner or later lose lambs. If you have vaccinated against tetanus and the other clostridial diseases with Covexin-8 (as per label) then the lamb will be protected. If not, you should administer 300 to 500 units of tetanus antitoxin to the lamb at docking.

CASTRATION

Castration also can be done early, as soon as the testicles have descended into the scrotum.

Emasculator. An Emasculator can be used for castration, so there is no wound. This is important in late lambing and warm weather, because it leaves no opening to attract flies. The Emasculator is a pincer instrument that gives bloodless castration by crushing the spermatic cord and arteries when you clamp it onto them like pliers. There is no loss of blood, less pain and setback to the lamb's growth, and no danger of infection. Check to see that testicles have descended into scrotum, then

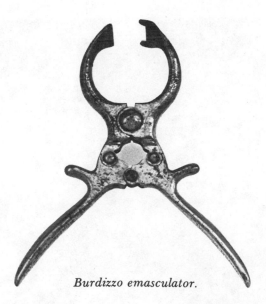

Burdizzo emasculator.

clamp the Emasculator onto the neck of the scrotum, where it joins the body, on each testicle cord separately. Because of the high cost of this well-made piece of equipment, you may not want to buy one for use on a few sheep. You might borrow it from a neighbor who has more animals or buy it in partnership with another sheep raiser.

After the Emasculator is used, the testicles will atrophy in about thirty to forty days.

The Emasculator can also be used for docking tails. Keep it disinfected. Push the skin toward the body, and crush between the joints. With a very young lamb, the tail is clamped in the instrument and pulled off. With older lambs, use a knife to cut it off just inside where it is clamped. The main artery is generally so crushed as to give quick coagulation of blood, and less bleeding than cutting off the tail with a knife alone. Douse the stump with disinfectant and spray with fly repellent if weather is warm.

Elastrator. The Elastrator also can be used for castration, when the lamb is about ten days old and when the testicles have descended into the scrotum. These special pliers stretch the rubber ring so you can pull the scrotum through it, being sure both testicles are down. When the pliers are removed, the ring tightens where it is applied, around the end of the scrotum where it attaches to the body, cutting off the blood supply so the testicles wither within twenty to thirty days. There is no internal hemorrhage or shock, and the risk of infection is slight. If you have problems with infection, douse the band with iodine after about a week. In hot weather, you can spray it with fly repellent.

Is castration necessary? There are reasons for not castrating—if you will be marketing the animal for meat at five or six months of age, or are thinking of keeping or selling it as a breeding ram.

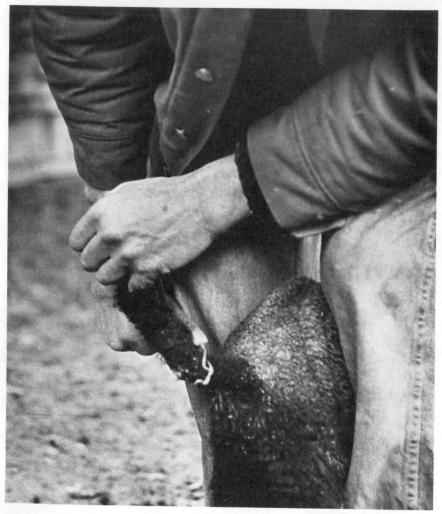

How to apply elastrator band to tail of three-day-old lamb. Special pliers stretch band for easy application.

While castrated lambs grow faster than ewe lambs, the uncastrated males will outgrow both of them, and the meat will be leaner. So, if you have early lambs and plan on selling the rams for meat at five months old (before breeding season), you can omit the castration. However, a packing house may penalize you $1 per animal or one cent per pound for not castrating, if that is your market. If you intend to keep the ram longer than six months before slaughter, castration is desirable.

Cryptorchid or short scrotum. There is still another approach, where the rubber Elastrator ring is used on the scrotum, but the testes are pushed back up into the body cavity. This sterilizes the animal, due to increased body heat. While the male hormones are still present to increase weight gain with more lean meat, the

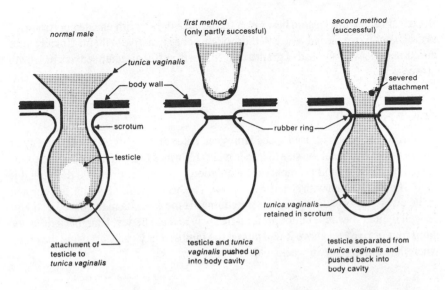

Methods used to induce cryptorchidism. (*Shepherd* magazine, December, 1973.)

animal shows little or no sex activity. This method is used at about four weeks of age, and the animal is called a cryptorchid (meaning "hidden testicles"). Extensive tests in Australia have shown such animals gain weight faster, get to market faster, and have less fat and more lean meat than either castrated or uncastrated males.

Although the body heat results in functional sterilization, do not use this method to create "teaser" rams. After an extended period of time, some develop a Sertoli cell tumor in the retained testicle(s) which produces abnormal amounts of female hormone that can actually cause feminization.

FEEDING LAMBS AND EWES

A ewe with twins (or triplets) cannot consume enough grass to support herself and give milk for them to grow, so she will need hay and grain until they are weaned. Even ewes with single lambs should have supplemental feed, unless you have lambed late and have a lot of pasture.

A ewe with a single lamb should have approximately a pound of grain a day, while a ewe with twins should get 1 1/2 to 2 pounds a day, plus some hay. Lambs from heavy milking ewes can gain up to 70 percent more during the nursing period than those from poor milkers. Lambs from good milkers will double their birth weight in two weeks.

In addition to their mother's milk, and the grass they start to nibble at about ten days old, the growing lambs need grain and hay in their own feeder, called a "creep." Start creep feeding early, as young lambs will investigate and use the creep more quickly than older ones, and it helps to establish their rumen function.

Grain and leafy hay are best given fresh twice daily, with uneaten portions fed to the ewes. To get the lambs to eat, the food must be attractive to them. Unless lambs are eating at least 3/4 pound of grain a day, they will suffer an acute setback in growth at weaning.

CREEP FEEDING

The creep is an enclosed space lambs can enter and eat all they want, but ewes cannot enter because of the size (8 inches wide, 15 inches high) of the doors and openings. The creep should be sheltered, with good fresh water provided daily, and it should be well bedded with clean hay or straw. The heavy stems of alfalfa, left uneaten in the ewes' hay rack, are good creep bedding. If the creep is in the barn, it should be well lighted, because lambs prefer it that way and eat better. Hanging a reflector lamp four or five feet above it will attract the lambs. They can start using the creep when they are about two weeks old.

MILK AND CREEP FEEDING FOR YOUNG LAMBS

No matter how well the lambs are taking to their creep feeding, they will still need a supplemental milk diet for long enough to allow their stomachs to develop so that their needs can be supplied by grain and forage. At birth, the lamb's rumen and reticulum have no microorganisms and are not capable of functioning like an adult's. Bacteria begin to populate the rumen shortly after birth, but it takes several weeks before a stable microbial population is established that is capable of efficient digestion. The late Dr. Beck explained it as follows.

Anatomically, all four stomachs are present in the lamb at birth, but only the fourth or true stomach, called the abomasum, is functional. The other three compartments develop as a result of stimulation by roughage; hence, the rate at which they develop depends on your feeding program. A lamb restricted to a milk nursing diet will develop the various stomach compartments at a slower rate than lambs started on creep and hay at, say, ten days of age. As a general rule, these compartments are "turned on" at about three to six weeks of age.

In other words, the early introduction of the creep feeding is important, especially if you wish to wean your lambs early, as is necessary in an accelerated lambing program when you intend lambing more frequently than the customary once a year.

Feed stores sell special "lamb creep feed," and there are lamb-feeding formulations that offer special advantages. Agway's High Energy Lamb Pellets (H.E.L.P.), for instance, is a popular one. It is made with fish meal for optimum quality, a 2-to-1 calcium-to-phosphorus ratio to help avoid urinary calculi problems, and vitamins and mineral fortification. It is available with either chloro-tetracycline to control enterotoxemia, or with Bovatec, which helps reduce coccidiosis. Since enterotoxemia (Overeating Disease) can be avoided much better by vaccination than by an overuse of antibiotics, the Bovatec form of pellets seems more sensible.

If you elect to feed grain and hay, but not pellets, then you can help prevent coccidiosis by the use of Decox, adding two pounds of Decox to fifty pounds of trace mineral salt.

In feeding grain, start with a mixture of crushed whole grains, plus some of whatever the mothers eat. The lambs always prefer the same grain that the ewes get, so sprinkle some of it on top of whatever creep mixture you use. The USDA recommends that if a commercial creep ration is not used, you prepare a mixture of the following percentages: 60 percent corn, 20 percent oats, 10 percent bran, and 10 percent soybean meal, with about 1 percent bonemeal and 1 percent mineralized salt added. This mixture can be coarse ground at first, then fed whole later. Since ewes milk heavily for only three to six weeks after lambing, it is urgent that the lambs be well adjusted to getting a good amount of their nutrition from creep feed and good quality hay.

WEANING

At weaning time, the lambs will adjust better if the ewes are removed, leaving the lambs in familiar surroundings. Weaning can be done gradually by putting the ewes in a different pasture during the day, and then returning them for the night. This has the advantage of keeping them from calling to each other and disturbing your sleep.

Ewes should have their grain diminished and then withdrawn completely at least five days before weaning, so that their milk supply will dwindle accordingly (to lessen the incidence of mastitis).

FORWARD CREEP GRAZING

If you have several pastures and rotate their use, you can give the growing lambs the benefit of the best grass by allowing them into fresh pasture ahead of the ewes, through creep-type openings in the fence. This will save creep grain amounts, and the lambs will grow faster and will have fewer internal parasites.

WORMING

Lambs are much more susceptible to worm infestation than adults because sheep, like many other species, develop a degree of resistance to worm infestation over a period of time. Lambs should be wormed at weaning time, using a safe wormer such as Lavamisole, Ivomec, Panacur, or TBZ. Read label directions for proper dosage, and note withdrawal times for lambs going to the locker. Lambs on lush, heavily stocked pastures or over-grazed pastures may need worming before weaning, and then again when they are separated from the ewes and placed on clean pasture. Worm populations thrive where warmth and rainfall (or irrigation) are sufficient to promote maximal grass growth. In some circumstances it may be necessary to worm lambs every three to four weeks.

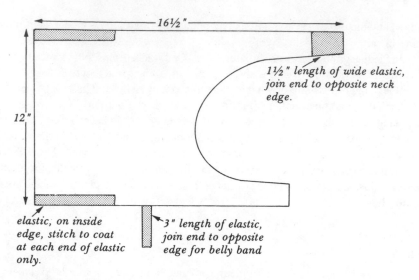

Here's a lamb coat for very cold weather, made from canvas or duck.

LAMB COATS

Newborn lambs are very susceptible to chilling because of their large skin area for their small body weight, and they are born without the fat covering under the skin that serves as a natural insulation against cold and chilling. Our northwest area has a moderate climate, but a long rainy season. Sheep don't mind a light rain. It usually takes a heavy rain, plus wind, to get them into the barn. This can create a problem when there are small lambs. If you keep the small ones penned for the first three days, they are quite hardy after that. In the Midwest, lambs are sometimes born during sub-zero weather. In extreme cold, most of the energy (calories) consumed by the lamb may be exhausted simply in maintaining body temperature, with no excess available for growth or for production of the protective fat layer under the skin.

Ewes can be encouraged to take their lambs to shelter during hard rain, by "facing" or trimming the wool from the ewe's face. The use of a lamb coat greatly reduces the heat loss, allowing the lamb to direct the energy it consumes toward growth and fat production. A canvas or duck or denim lamb coat can be made, using the pattern above. There is a new plastic disposable lamb coat available, made in perforated rolls which allow one to be torn off as needed. A less expensive disposable lamb coat, very similar to the commercial coats, can be made from white plastic garbage pail liners. By folding the bag lengthwise, and using the cardboard pattern shown, two lamb coats can be made from a single bag. My veterinarian friend uses these homemade coats exclusively because they are tear resistant, but not so strong that the lamb cannot walk out of it in case it gets caught or snagged. Since they are completely open at the rear and the bottom, they are very sanitary and they do not confuse or frighten the ewe.

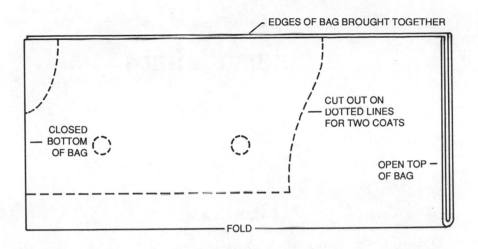

EDGES OF BAG BROUGHT TOGETHER

CUT OUT ON
DOTTED LINES
FOR TWO COATS

CLOSED
BOTTOM
OF BAG

OPEN TOP
OF BAG

FOLD

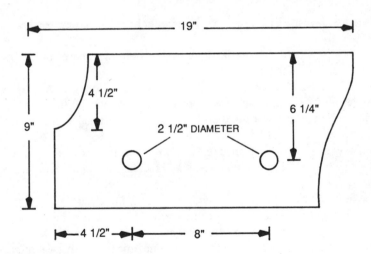

19"

4 1/2"

6 1/4"

9"

2 1/2" DIAMETER

4 1/2"

8"

Pattern for two plastic lamb coats. Bag brand used was HeftyR brand Kitchen Garbage Bag, 13 gallon size.

CHAPTER 10

Orphan Lambs

LAMBS & EWES

ORPHAN ("BUMMER") LAMBS can result from the death of the ewe, abandonment, rejection, or loss of milk production before the lamb has reached weaning age. A ewe may disown one or all of her lambs, sometimes for reasons known only to her. The following are the most common reasons:

- The ewe may have a painful or sensitive udder because of overabundance of milk, or mastitis.
- She may have delivered one lamb in one location, then moved and delivered the other, forgetting about the first.
- Some ewes cannot count to "two." They may be willing to accept twins, but as long as they have one, they are happy and do not seek out the other.
- The lamb may have wandered off before the ewe has had a chance to lick it off and become bonded to it.
- She may have sore or chapped teats, or the lamb may have sharp teeth.
- Because of a difficult lambing, she may be exhausted and not interested in her lamb.
- The lamb may be chilled and then be abandoned as dead.
- New mother syndrome: If a young "first-time" ewe, she may be nervous, flighty, confused, or just frightened of the lamb.
- Swapping lambs: If two ewes lamb at the same time in close proximity, occasionally one ewe will adopt and bond to the other's lamb, and the second ewe will reject the first ewe's lamb.
- Nervous shepherd syndrome: Grafting a lamb onto a ewe that has just dropped a single, only to find that she subsequently drops twins or triplets and cannot feed the mob.

PERSUADING THE EWE

This will severely test your patience and ingenuity! If a ewe has a single lamb which she rejects, you have double trouble, because you not only have a hungry lamb but an increasingly uncomfortable mother as well. You want to get her to accept her lamb. If she rejects one of a pair of twins, either you can convince her to accept it, or you can attempt to graft it onto another ewe who has lost her lamb, or has only a single.

Your first consideration is the urgent need for the lamb to receive colostrum, so either roll the ewe on her side and put the lamb's nose against her teat to get it to nurse, or milk the ewe and feed the lamb with a bottle (or stomach tube for a weak lamb). Colostrum contains the antibodies that protect the lamb from all the germs in the environment until its immune system can build its own protection. Lambs can survive nutritionally without colostrum, but it is very difficult for them to survive without the disease-protecting antibodies in the colostrum. When your ewes have been vaccinated with Covexin-8, they are passing along immunity to tetanus and other clostridial diseases, through the colostrum. Try to provide the lamb with several nursings of the vital colostrum—either from its own mother or from another ewe.

In most cases the lamb is hungry and very cooperative. Tickling it under its tail helps to stimulate the sucking reflex. This first feeding gives you a little time to arrange a forced acceptance by the ewe. Do not leave a rejected lamb unattended with the ewe, as she may injure the lamb by stepping on it or butting it.

Should the ewe reject the lamb *after* it starts to nurse, not before, check her udder for sensitivity, and check the lamb's teeth. A little filing with an emery board can remedy sharp teeth. Don't file too much or the teeth will be sore and it won't nurse, which puts you right back where you started. Apply Bag Balm to the ewe's teats if they are sore or lacerated by sharp teeth. Keep her tied where the lamb can nurse until she accepts it.

There is scientific evidence suggesting that the vaginal stimulation during parturition plays a large role in the ewe's instinct to accept the lamb, which could explain why the success of grafting lambs is greater when attempted as close to the delivery time as possible. This could also explain why some "easy lambers" simply walk away from a newborn lamb as if its birth were just a minor occurrence.

Once a ewe rejects a lamb for any reason, it is hard to fool her into accepting it. All methods fall into two major categories: (1) the mental or "brainwashing" techniques in which you attempt to change their hard-headed opinion, or (2) the physical or "fool the sense of smell" method.

There are a number of things to try, such as:

1. Use fetal fluids from the ewe that the lamb is to be grafted to (either its mother or another ewe) and smear over the lamb. This is one of the most effective "tried and true" methods of grafting.
2. Rub the lamb with a little molasses water, to encourage the ewe to lick the lamb.
3. Use an "adoption coat" or "fostering coat" (see Sources) which is a cotton stockinette tubing applied like a lamb coat. When stretched over an accepted lamb for a few hours, it will absorb the smell and can then be turned inside-out and stretched over the lamb you wish to graft. (Shepherd's tip: If you have a heavy milking ewe with a single, slip a coat on her lamb in order to have a fostering coat ready to use if needed.)

4. Daub her nose, and the lamb's rear end, with a strong scent-masking agent such as U-Lam or Mother Up which are made for this purpose. Since the ewe identifies the lamb primarily by smelling its rear end, sometimes Mentholatum, vanilla, or even a nonscented room deodorant on her nose and the lamb's rear, will suffice.

5. If it is a case of the "new mother jitters" or the ewe is high-strung and not very tame, a tranquilizer will sometimes work wonders to calm her.

6. There is an old-timer's method of tying a dog near the pen. Its presence is supposed to foster the mothering instinct. I found this sometimes makes the ewe so fierce that she will butt the lamb if she can't reach the dog.

7. Another method, which is not actually as cruel as it sounds, is to flick the tips of the ewe's ears with a switch until she becomes so rattled that she urinates from the mental stress. She may then accept the lamb.

8. Immerse the lamb to be grafted (or owned by the ewe), as well as the lamb she accepts, in a saturated salt solution to even out the scent. Again, though, I worry that she may become confused and reject both.

FORCIBLE ACCEPTANCE

If all else fails in your fostering attempts, then it is time to get tough. One solution is to pen or tie the ewe in such a way that she cannot hurt the lamb, and it can nurse regularly in safety. You may need to tie her hind legs together temporarily, so she can't keep moving and thus prevent the lamb from nursing. Without the mother's guidance and encouragement, you may need to help the lamb nurse by holding the ewe and pushing the lamb to the right place.

If the ewe is a hard-core case, a ewe-stanchion could be necessary (see plans). A less elaborate one can be improvised in the corner of the lambing pen. Make sure that the ewe has room to lie down, and has plenty of hay and water in front of her. Use molasses in the water, as you would for any ewe who has just lambed. It may take from one to five days before the ewe is completely resigned to accepting the lamb.

Caution: Exercise care and judgement in the size of the lamb that you are attempting to graft. An orphan lamb that is one or two weeks of age may be so aggressive at nursing that it will frighten the ewe. Also, if there is a significant difference in age and size between two lambs placed on a ewe, the weaker lamb may not be able to compete with the larger lamb and will suffer restricted growth or, at times, may be starved out completely.

Twins. The most typical situation is the birth of twins and the rejection of just one of them. Spraying the rear end of both lambs with a confusing scent is the easiest thing to try at first, and most often it works.

If the ewe starts showing any hostility *at all* toward one of her twins, either by acting suspicious of one, or by talking with soft baby talk to one and with a grumpy sound to the other, don't wait until she starts butting it, but take positive action right away. The most reliably successful way is to tie her up. The sooner you stop her from comparing the smell of the two lambs, the sooner she will accept the reject. In any make-do tying arrangement, be sure she gets water often, for it may be difficult to leave it in front of her.

Some suggest that you tie up the ewe and leave the reject lamb with her, but take away her favorite, bringing it back only to nurse. We have not found that necessary, and I would hesitate to do it because I'd be so very frustrated if she accepted the reject and decided she didn't want the one I had been taking away.

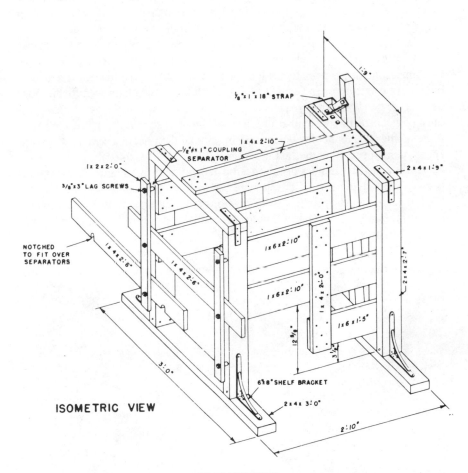

ISOMETRIC VIEW

LIST OF MATERIALS

Lumber:	Miscellaneous:
2 pcs. — 2 x 4 x 10'-0''	6 — ½'' x 1'' pipe couplings
1 pc. — 2 x 2 x 7'-0''	4 — 6'' x 8'' shelf brackets
2 pcs. — 1 x 6 x 8'-0''	1 — ⅛'' x 1'' x 4'-0'' steel strap
1 pc. — 1 x 4 x 12'-0''	1 — 1½'' x ⅛'' x 5'' steel strap
1 pc. — 1 x 2 x 4'-0''	4 — ¼'' x 2'' lag screws and washers
	10 — ⅜'' x 3'' lag screws and washers
	48 — 1¼'', No. 9 wood screws
	¾ lb. — 6d common nails
	¼ lb. — 12d common nails

In the meantime, if another ewe goes into labor and you think she may deliver only one lamb, you might choose to graft on the reject, for she may be more willing than the ewe who is all geared up to rejecting something.

GRAFTING AN ORPHAN ON A DIFFERENT EWE

Have a bucket of warm water ready and also an empty bucket. Have the rejected lamb nearby, and watch the lambing. If you are fortunate enough to catch the water bag, put its contents into the empty bucket; this makes everything much easier.

As the ewe delivers her own lamb, dunk the waiting reject into the water-bag liquid. Or, if you didn't catch that, into the warm water up to its head. Then rub the two lambs together, especially the tops of the head and the rear ends. Present them both to the ewe's nose, and usually she will lick them and claim them both. Don't neglect the newborn when you are working with the orphan—the new lamb's nose must be licked off by its mother or wiped off by you, so that it can breathe. Now, if the mother delivers twins, you may have to take the reject back. Dry it off, and keep trying to get its mother to take it (or bottle-feed it yourself).

If the substitute mother does appear to accept the grafted lamb, and that lamb is much older than her newborn, hobble the orphan's legs so it doesn't get up and run around too much at first. Let the newborn lamb have the first chance to nurse. If your orphan is a few days old, it doesn't really need the colostrum, and should not get too much of it at one time.

Targhee ewe with newborn lamb. (Mt. Haggin Livestock Inc., Anaconda, MT)

Actually, to do this trick, the orphan should be less than a week old, as an older one would surely cheat the new lamb out of its share of the milk. In any event, both lambs will have to be supervised carefully.

Water Bag Stockpile. One worthwhile practice is to save the water bag from a ewe and freeze it in pint quantities. You can thaw this and pour it over a ewe's nose and onto the lamb you want to make her accept. Not always successful, but worth trying.

GIVING ORPHAN TO EWE WHO HAS LOST HER LAMB

When you find a ewe who has delivered a dead lamb, and you have a young orphan who needs a mother, dunk the lamb in warm water containing a little bit of salt and some molasses. Dip your hand in the warm water and wet its head. By the time she licks off the salt and molasses, she usually has adopted the lamb. When it is a lamb that is several days old, and does not need the colostrum as much as a newborn, this gives you an opportunity to milk out and freeze some of the valuable fluid.

In all this talk about grafting an orphan onto a ewe, I've not mentioned the old way of the "dead lamb's skin." In that method, if a lamb were born dead, or died after birth, it was skinned and the skin fastened like a coat over the orphan. Skinning a dead lamb is not simple unless you are already adept at it. The process is messy and unsanitary, since you may not know why the lamb is dead, and could be transferring germs and disease.

Another less messy method is to rub a damp towel over the dead lamb, then rub the towel on the orphan. Before doing this, wash the orphan with warm water, giving special attention to washing the rear end which is the first place the ewe checks in determining whether the lamb is her own.

I remember a postage stamp issued some years ago, showing a ewe with a lamb. She appeared to be sniffing its head. Sheep raisers laughed, as it was a very untypical end for her to be sniffing.

The fewer sheep you raise, the less chance there is that another ewe will be lambing about the time you need a substitute mother. So if its mother has died, has no milk, has been incapacitated by pregnancy disease or calcium deficiency, or completely refuses to accept her baby, you have a bottle lamb.

BOTTLE LAMB

This is one of the greatest pleasures (and biggest headaches) of sheep raising.

The lamb's first need is to have its nose mucus wiped off so it can breathe. Even if the ewe is weakened by a hard labor and/or has no milk, she should be allowed to clean the lamb as much as she will; if unable to nurse, she will still claim it, and even as a bottle lamb it can stay with her. If the ewe does not lick off its nose, you wipe it off, then dry the lamb and put iodine on its navel at once.

COLOSTRUM

Now, it needs some real colostrum (ewe's first milk) if possible, either from its mother who may have rejected it or is too weak to stand up (roll her over and help the lamb), from another newly lambed ewe, or defrosted from the freezer if you have it. Cow or goat colostrum are the next best substitute for ewe colostrum, although there is a commercial preparation of colostrum powder that is useful (see below).

The new Colostryx is a milk whey antibody product for lambs, and transfers a certain amount of immunity to the newborn when mixed with milk-replacer (or diluted canned milk or cow milk) for the first day.

If this is your first year with sheep and lambs, you will not have frozen colostrum on hand. While the commercial colostrum powder is manufactured for calves, it is still better than none at all. For orphan lambs, the very best thing is to give 1 or 2 ounces of another ewe's colostrum for the specific local antibodies. Then mix 1 ounce colostrum powder with 1 cup warm water, for the first twelve to eighteen hours of feeding. After that, 1 ounce Colostrum powder with 2 cups warm water for the next day. Then, 1 ounce Colostrum powder can be mixed with a quart of lamb milk-replacer for feeding, or with canned milk diluted with 1/2 water. After that, just use the lamb milk-replacer. For the first forty-eight hours you can feed the lamb every three hours, with no more than 1/2 cup per feeding, at most. On the third day, you can add a child's vitamin, crushed.

It is possible, but difficult, to raise a colostrum-deprived lamb. In colostrum the lamb receives antibodies against common environmental pathogens, which protect it during the first few weeks of life until its immune system can begin producing them. The lamb is too young to receive a Covexin-8 vaccination (although some people do it) and you will need to administer antiserum in order to protect it temporarily against enterotoxemia and tetanus. Your veterinarian can supply this.

Cow colostrum can be a valuable nutritional substitute for ewe colostrum. A pregnant cow can be vaccinated with sheep vaccine Covexin-8, several times, and the first two milkings of colostrum will be high in antitoxins and give lambs good protection. (This would give an enormous quantity of colostrum for freezing.) The lambs can still be given vaccination at six to nine weeks, which will then protect them up to twenty-four weeks. The best time to inject the cow would be with 5 ml six weeks prior to calving, with a booster of 5 ml two weeks before calving.

The advantages of ewe (or even cow) colostrum from your own farm is that its antibodies are "farm specific," to protect the lamb against any organisms the ewe (or cow) was exposed to on your farm.

Emergency Newborn Lamb Milk Formula

While this is in no way a satisfactory substitute for colostrum, it can be fed for the first two days, rather than just starting out with milk-replacer.

26 ounces milk (1/2 canned milk, 1/2 water)	1 Tbsp. castor oil (or cod liver oil)
	1 Tbsp. glucose or sugar

1 beaten egg yolk

Mix well, and give about 2 ounces at a time the first day, allowing from two to three hours between feedings. Use a baby bottle and enlarge the nipple hole to about the size of the head of a pin, or make a small X hole. Lamb nipples are larger, so use one of these when the lamb is older.

On the second day, increase the feedings of the formula to 3 ounces at a time, or 4 ounces for a large, hungry lamb, two to three hours apart. On the third day the formula can be made without the egg yolk and sugar, and the oil can be reduced to 1 teaspoon per 26 ounces of milk.

After the third day, you can gradually change to lamb milk-replacer. Do not use milk-replacer that is formulated for calves; it is too low in fat and protein. A feed store can special order the lamb milk-replacer if they do not have it in stock.

Goat's milk is also a good lamb food. With that available, you wouldn't need the powdered milk-replacer. Don't overfeed at any time, it is better to underfeed than to have a sick lamb. A bottle lamb is more subject to infections than a lamb on the ewe's milk, so keep bottle and nipples clean.

If the lamb's droppings get runny (diarrhea is called "scours"), use Pepto Bismol immediately. Give several teaspoons for a lamb a week old or younger. Give more for a larger lamb, and cut back on the amount of milk given, as well as diluting it more. This condition can be caused by bacteria, but is more often just the result of overfeeding.

Keep milk refrigerated, warming it at feeding time, and keep milk containers clean, to prevent the kinds of scours caused by bacteria.

LAMB MILK-REPLACER

A comparison between the contents of cow's milk and sheep's milk will show the difference there will be in the replacer milk made for their offspring:

	Grams of Protein	Grams of Fat	Grams of Calcium
Ewe milk 1 cup	11.0	13.0	0.413
Cow milk 1 cup	7.0	7.8	0.236

From *Table of Food Values* by Alice V. Bradley.

Ewe's milk has almost twice as much fat as cow's milk, and the lamb milk-replacer has about twice the fat content, on a solids basis, as calf milk-replacer. A newborn lamb is a small creature with a high ratio of body surface compared to its heat-generating capacity (more so than a calf). The fat provides this heat generation. Nature provided ewe milk with high fat content for a good reason.

Carl Hirschinger of the University of Wisconsin says to look for a lamb milk-replacer that contains 30 percent fat and 24 percent protein, on a dry matter basis, and no more than 25 percent lactose (milk sugar). High lactose levels can cause diarrhea and bloat. Whey products are high in lactose, and should be delactosed, or used sparingly in the replacer. Hirschinger recommends diluting replacer to a minimum of 15 to 20 percent dry matter for the first week.

Most lamb milk-replacers are medicated for prevention of many common lamb ailments, and fortified with vitamin A, which is very high in natural ewe's milk, as well as with vitamin D and vitamin E, which is useful in the prevention of stiff lamb disease. The milk-replacers also have necessary minerals.

We mix the replacer powder in a blender, make enough for two days, and store it in the refrigerator. We mix ours double strength, using only half as much water as the recipe calls for, and then dilute it with an equal amount of hot water for bottle feeding.

One suggestion I would make for mixing milk-replacer: Substitute *limewater* (see recipe below) for about one-half of the water in the concentrated, double-strength formula. Limewater is a sedative, antacid, astringent, makes milk more easily digested, and reduces the tendency to scours.

To make limewater, use ordinary slaked lime purchased from a feed store. If you have trouble finding the slaked lime, buy it as calcium hydroxide in the drugstore. You need only a very small bottle of it. The 0.4-ounce bottle will last for a lamb's three months of feedings. Some drugstores sell the plain limewater already made up (this is *not* the juice of limes) because it is used in some baby formulas.

Limewater Recipe

1 teaspoon slaked lime
1 gallon of water

Add the lime to the water. Shake it several times during the day, then let it stand until it is clear. Drain off the clear liquid and use as 1/4 of the water in the milk-replacer mixture. (Or, you can make up half of the limewater recipe at a time with 1/2 teaspoon lime to 2 quarts of water.)

A slow-flowing nipple is best, for then the lamb cannot gulp its milk and choke or get sick. A hole about the size of a pin head is about right for the first couple of weeks and can be enlarged just a little as needed. By the time the lamb is a month old, you may want to change to a lamb nipple, which fits over the top of a soft-drink bottle.

As the lamb gets older, it drinks more and gets correspondingly louder at mealtime, bouncing its bottle like a punching bag and making a real pest of itself. Sooner or later the inevitable happens; it takes a mighty swig at the bottle and pulls off the nipple, splashing everyone with milk and creating a threat of disaster—the possibility that it will not only chew but swallow the removed nipple before you can take it away.

The lambs come and get it at this New Zealand experiment in raising large numbers of lambs on bottles. Note each lamb has its own bottle. (D.H.B. MacQueen, photographer, Raukura Agricultural Research Center)

BLOATED BOTTLE LAMB

This is an infrequent situation, but it can happen if the lamb is overfed or if it drinks too fast (nipple hole too large). Cut back on the amount of milk being given, and give one small feeding of two ounces of milk containing about one tablespoon (for lamb under one month) or two tablespoons (for lamb over one month) of human antacid medication with simethacone. This is similar to a veterinary bloat remedy called *Bloat Guard*, containing methyl silicone. If the lamb will not take it in the bottle, give with spoon, carefully.

Finnsheep sextuplets. Four of these were raised as bottle lambs. (USDA, Beltsville)

SUGGESTED FEEDING SCHEDULE FOR ORPHAN LAMB

Age	Amount
1–2 days	2–3 ounces, six times a day, approximately (with colostrum or formula with Colostrum Starter powder)
3–4 days	3–5 ounces, six times a day (gradually changing over to lamb milk–replacer)
5–14 days	4–6 ounces, four times a day, and start with leafy alfalfa and crushed grain or pelleted creep feed
15–21 days	6–8 ounces, four times a day, along with grain and hay
22–35 days	Slowly change to one pint, given three times a day. After lamb is three months old, feed whole grain and alfalfa, or pelleted alfalfa containing 25 percent grain—but change ration *very* gradually.

For lambs nursing ewes, nature automatically regulates the amount of milk that they receive per feeding—small amounts, but often. It is very important to control the volume of milk fed per feeding to bottle lambs. It is so tempting to overfeed a bummer lamb—it is so cute, and soon learns to beg in an irresistible manner. A yellow semi-pasty diarrhea is the first sign of overfeeding. If this occurs, substitute plain water or an oral electrolyte solution such as Gatorade for one feeding, because the lamb needs the fluid but not the nutrients. Reduce the volume fed until the condition clears. If the droppings become more runny, give it Pepto Bismol in the quantity mentioned under "Colostrum" heading.

Overfeeding is more common during the first week or two of life than it is later on. My veterinarian routinely starts his newborn orphans on lamb milk-replacer (after colostrum) that is prepared with twice the volume of water than is recommended on the label. (This is the only exception to his "always-read-and-follow-the-label" sermon.) He then gradually increases the concentration of milk powder in the solution until it is at full strength by the time the lamb is a week or so of age. At first sign of the "yellow goo's," he reduces the concentration slightly for a day or so, and then gradually brings it back.

As orphan bummer lambs get older, their need for water increases, especially if they are beginning to eat grain from the creep feeder. If they have not yet learned to drink from the water tank, they will attempt to quench their thirst with milk, which is the equivalent to you attempting to quench you thirst on a hot day with a milk-shake! Substitute an occasional feeding with plain water, or dilute their regular feeding with more water and increase the amount given. Judging the need for water in a bummer lamb will require experience and the development of a shepherd's sixth sense. When feeding bummers, common sense and observation are your best allies.

The true bummer is a lamb whose mother either dries up or doesn't have enough milk, and the lamb is forced to sneak or "bum" off other ewes. If this occurs prior to three to four weeks of age, the lamb may be skinny, lose weight, or even starve. If the lamb is big enough or smart enough, it will figure out how to "bum" from the other ewes in the flock without getting caught. It usually sneaks up behind a ewe just after grain is fed and the ewe's attention is focused on competing for her share, or when her head is thrust into the hay feeder. Most ewes become less protective of their lamb, and hence less particular about who may be nursing, as the lamb gets older. Bummer lambs seem to seek out these ewes and can often be seen nursing from behind, between the legs.

HOT FLASHES

If you happen to be holding a young lamb on your lap after bottle feeding (which is sometimes irresistible), you may notice that it suddenly feels very hot or flushed, approximately five to ten minutes after feeding. This "hot flash" is not actually a sudden increase in body temperature, but rather an acute dilation of the capillaries of the skin, which releases a short burst of body heat. These hot flashes usually last only a minute or so. Do not become alarmed, as it is a known physiological phenomenon of sheep (and cats). The mechanisms and reasons for it are poorly understood.

ORPHAN FEEDING, CAFETERIA STYLE

If you raise Finnsheep or another breed that has multiple births, or a flock large enough to have quite a few orphans, there are several commercially made multiple-nipple arrangements for feeding them, such as Lam-Bar and Lambsaver. The lambs

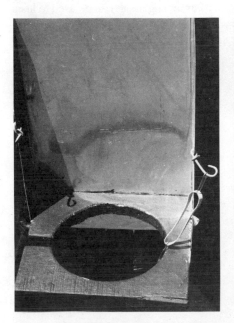 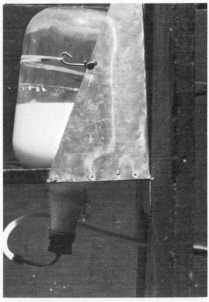

This lamb feeding arrangement was designed by D.T. Torell of the Hopland Experiment Station. The *Lam-Bar* nipples used with it can be purchased from Sheepman Supply (see Sources).

should first be taught to nurse from a bottle, on warmed milk-replacer, then changed to the feeder.

With this system, milk is always before the lambs, and they can suck it out as they want it. The milk formula is usually fed cold to reduce the chance of overeating, and to reduce bacterial contamination when it is left standing all day. The lamb milk-feeder should be cleaned and disinfected and supplied with fresh milk daily. For such use, the replacer should be one that stays in suspension well.

D.T. Torell, at the University of California's Hopland Experiment Station, did tests showing that the addition of one milliliter of Formalin (37 percent solution of formaldehyde in water, available in small quantities from most pharmacists) to each gallon of liquid milk-replacer being fed at barn temperature, would keep the milk free of bacteria for several days without having any adverse effect on the lambs. While it is still preferable to have fresh milk daily, the tiny amount of Formalin would eliminate the chore of such careful disinfecting of the feeding equipment. The one milliliter of Formalin is the same as one cubic centimeter on your hypodermic syringe. You also might measure it with a one cubic centimeter insulin syringe.

Here are illustrations of the Hopland Lamb Nursing System, developed by Mr. Torell. It is a gallon jug suspended upside down, with a No. 6 rubber stopper from which a 1/4-inch-diameter plastic tube protrudes. The tube then leads to another No. 6 stopper holding a nipple which is inserted through the board to the lamb area or pen. The nipple is available for nursing at all times.

The hole in the board for the nipple must be located 1 or 2 inches above the level of the jug neck. A strip of inner tube or other strong elastic can be tied on each side to screw-eyes, to hold the jug in place. A similar arrangement can be used to keep the stopper and nipple firmly in the hole while the lamb nurses. One of these Hopland Nursing Systems will feed three or four lambs, using lamb milk-replacer fed at barn temperature.

CHAPTER 11

Lamb Problems

WEAK LAMB AT BIRTH

A LAMB WEAKENED by a protracted or difficult birth may be suffering from anoxia (lack of oxygen) or have fluid in its lungs. The first few minutes are critical. If it gurgles with the first breaths or has trouble breathing, dry off the nose, grasp the lamb firmly by the rear legs, and swing it upward vertically in a gentle arc, catching the lamb momentarily on the return end of the upswing with the free hand so the lamb is stopped abruptly with the head up in a vertical position. This does two things: (1) centrifugal force aids the movement of the fluid from the lungs, and (2) the weight of the viscera presses on the diaphragm, causing a forced expiration. When you catch it vertically on the upswing, the weight of the viscera falls in the opposite direction, causing a forced inspiration. Normally two or three "swings" will get things going. Be sure that you have a firm grasp on the lamb (the lamb will be very slick) and that there are no obstructions in the path of your swing.

If the heart is beating, but the lamb is still not breathing, artificial respiration is mandatory. Grasp the lamb by the nose so that your thumb and fingers are slightly above the surface of the lamb's nostrils. Inflate the lungs by blowing GENTLY into the lamb's nostrils until you see the chest expand. Release the pressure and gently press on the lamb's chest to express the air. Repeat the procedure until it begins to breathe. Exercise caution, don't blow as though blowing up a balloon, its lungs are quite small and can be ruptured by too much pressure. If your attempts are still unsuccessful, sometimes a cold water shock treatment will do the trick. Dunk the lamb in cold water, such as a drinking trough. The shock may cause the lamb to gasp and start to breathe. Sometimes a finger inserted gently down the throat will stimulate the coughing reflex and get things going. Then, make sure the lamb is warmed and get it to nurse.

Avoid excessive heating and unnecessary use of a heat lamp if the lamb's temperature is normal (102 degrees) because too much of a temperature differential will predispose it to pneumonia. However, if a newborn lamb is so cold from exposure that its mouth and tongue feel cold or cool to the touch, there is no time to warm it with a heat lamp. You must apply external heat in this case, because the lamb has lost its ability to maintain and control its body temperature. The quickest method to warm a chilled lamb is to immerse it in hot water. The water should be comfortable to the touch to begin with, then heated to 110 to 115 degrees F. over a period of five to ten minutes. Move its legs around in the water to increase circulation. When its body temperature reaches 100 degrees, the mouth and tongue will again feel warm. Keep the lamb in the water until the temperature is near 102 degrees, then rub it dry, give warm milk if it will suck, and wrap it in a blanket until it begins to regain its strength. Heated water has advantages over a heat lamp, as it is faster, easier to control the temperature and does not tend to dehydrate the lamb. Until recently lambs were immersed in hot water, but some deaths attributed to chilling were found to be due to shock, which does not happen when the water is warmed more gradually.

Soaking the newborn in water removes its natural odor, causing more danger of ewe rejection. And sometimes hot water soak is not quickly possible, and it does require one person to hold the head above water at all times. An alternative is hot air. Use a small box with an opening so the lamb's head can be outside to breathe fresh unheated air, and a small opening through which a hair dryer nozzle can be directed. Medium or low heat may give the temperature needed. Test with your hand to be sure it is warm enough, but not too hot. The lamb needs to be turned and rubbed, and its legs exercised, occasionally. When its body temperature is 100 degrees, see if it will suck, or else use a stomach tube. Maintain heat until its temperature is normal. A severely chilled lamb may need four hours of heat to return to normal temperature.

If you have no electricity in the lamb pen, put the lamb in a small box with a hot water bottle or a plastic jug filled with hot water, to warm it until it is dried off. This would not be sufficient for an extremely chilled lamb. When using a hot water bottle (not hot enough to burn!), apply the heat first to the belly where it is most needed.

You must stomach-tube with colostrum if it has not regained its sucking instinct, as the lamb is in dire need of energy. If you do not have colostrum, then milk-replacer, cow's milk, diluted canned milk, electrolyte solution (or Gatorade), or water with a small amount of white corn syrup will suffice. Feed about 1 ounce.

It is not completely necessary that the very first feeding be colostrum, only that the lamb receive colostrum in subsequent feedings during the first few hours of life. Its ability to absorb the antibodies in the colostrum is a straight-line decrease from time of birth to approximately sixteen hours of life. After that, it has lost its ability to absorb the life-protecting antibodies, no matter how much colostrum you feed. The important thing is to get the lamb going first, then worry about the secondary necessities such as pneumonia protection. A small shot of Pen-Strep immediately, followed by Naselgen-IP vaccination at ten to eighteen hours of life will help protect it from pneumonia.

When a lamb a few hours old is crying continuously or has a cold mouth, it is not nursing. The ewe may not have milk (or you may have forgotten to strip her teats to unplug her), or the lamb may not have found the udder. Check to determine the problem.

A weak lamb that has not been able to stand up to try to nurse within a half hour will need help. Hold it up to the ewe if she will stand still, or put the ewe down and hold the lamb to nurse. Use this same procedure for a stronger lamb, if it has not located the right place and begun to nurse within one hour after birth. I would seldom wait that long.

For a very weak lamb, you may have to give the first feeding from a baby bottle with the nipple hole enlarged to about the size of a pin head. Use two ounces of the ewe's colostrum, warm, to give it strength. Do not force the lamb, if it has no sucking impulse, or the milk will go into its lungs and cause death. Try the dextrose injection (described in next section) which is easy, then wait for a half-hour to see if this gives it energy and the desire to suck. If not, then try the stomach-tube feeding method described later in this chapter.

Colostrum is more important to lambs than to other animals, because lambs get no protection from antibodies transferred to them while they are still in the uterus. They are completely dependent on colostrum to protect them against the germs they encounter as soon as they are born.

The gut of the newborn lamb does not break down the proteins in colostrum, but absorbs them unchanged; thus the antibodies remain intact and are immediately usable. This process of antibody uptake decreases and is lost after sixteen hours, so it is urgent that the lamb gets its colostrum feedings soon.

If the ewe has no milk and you have no frozen colostrum, commercial Colostrum Powder, nor another newly lambed ewe to swipe it from, see the "emergency colostrum" formula in Chapter 10. Proceed as for an orphan lamb, but keep it with its mother if she is attentive, for the lamb will be happier and healthier with a sheep-mama to follow around. The new Colostrum Powder on the market sounds like a good thing to have on hand in case no frozen colostrum is available. (See Sources.)

DEXTROSE INJECTION REVIVAL, FOR A VERY WEAK NEWBORN LAMB

Dextrose injections can be given if the lamb cannot suck and you do not have a stomach tube or are not comfortable with using one.

You can give a real shot of energy with a dextrose injection, using 50 ml of a 5 percent dextrose solution in saline, warmed to body temperature. This must be purchased from a veterinarian or animal health supplier, as you cannot make your own.

Warm the solution to body temperature, and inject it subcutaneously (under the skin) in divided doses of 5 to 10 ml per injection site. Any area of loose skin in the neck area or behind the armpits at the point of the elbow on the chest wall, are injection sites. The injections will make unsightly bumps under the skin, but don't worry, they will resolve very rapidly. Never make injections directly into the armpit, as it can be irritating because of poor circulation in this area. Use a sterile disposable 18- or 20-gauge needle. Sterilize the top of the vial with alcohol, wipe dry, and insert a disposable sterile needle into the stopper to fill the syringe. Leave this needle in the vial and use a separate needle to make the injections, to avoid contaminating the glucose solution. Sterility is important. Even the slightest contamination will grow very rapidly in glucose, ruining the vial for further use. Always refrigerate the glucose solution after opening, and do not use it if it becomes cloudy. Calcium solutions containing dextrose, or 50 percent dextrose should not be used for this purpose. They are really intended for intravenous use, are very irritating to tissue, and the calcium solution could cause heart blockage and death in an animal that is not calcium deficient.

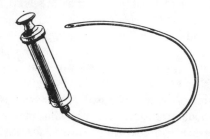

One arrangement for emergency stomach tube feeding.

WEAK NEWBORN LAMB—STOMACH TUBE EMERGENCY FEEDING

The Sheepman Supply Company (see Sources) has a device called a "Lamb Reviver" which you can order by mail, for the stomach tube feeding of a severely weak lamb with no sucking impulse. If you need one quickly and there is no time to order, get a male catheter tube from the drugstore, and use it with a rubber ear-syringe or a 60-cc hypodermic syringe for a direct feeding into the lamb's stomach. The tube should be about 14 to 16 inches long. (Check length by holding against the lamb.) Before inserting to inject the milk, disconnect the tube from the milk-filled syringe, so that you can determine that the tube is actually in the stomach and not in the lungs. An injection into the lungs would kill the lamb. The tube should be kept in warm sterile solution; when wet it slips in more easily.

In slipping the tube into the stomach, you should actually be able to feel the tube as it goes down, if you put your thumb and finger along the outside of the neck and pass the tube with the other hand. If the tube is incorrectly passed in the trachea, you can't feel the tube going through the neck. A tube into the lungs will usually elicit a cough, but I would not depend on that as a sign. When the tube is in what you think is the correct position, hold a wet finger at the protruding end—if the finger feels cool from moving air, the tube is in the lungs instead of the stomach, so remove it and try again.

Another way to check for proper position is to blow gently on the tube. If it is in the stomach, you can see the lamb's abdomen expand. Releasing the pressure will allow the air to escape and you will see the abdomen flatten again. If the tube is in the lungs, the air will escape past the tube and up the trachea without this "ballooning" effect.

It is easier for two people to operate the stomach tube, but it is possible with one person, if the syringe is filled in advance with 2 ounces of warm colostrum (or warmed canned milk, undiluted, for this feeding only) and placed within reach. The following is the procedure for one person.

Position of lamb. On a table, with its feet toward you, hold the lamb's body with your left forearm, making a straight line between the lamb's head and neck and back. Use fingers of left hand to open the lamb's mouth to insert the tube, which should be sterile and warm, if possible. Insert the tube slowly over the lamb's tongue, back into its throat, giving it time to swallow, and push the tube down its neck and into the stomach (having checked the tube length previously, so you know about how much of it should stick out), when the end is in the stomach area. The average insertion distance is 11 or 12 inches. You cannot insert it too far, but it is important to insert it far enough.

When you have confirmed its correct position with the wet-finger or blowing test, insert the end of the catheter tube into the syringe filled with warmed milk and slowly squeeze the milk into the lamb's stomach. Withdraw the tube quickly, so it will not drip into the lungs on the way out. Helene Lund, of *Shepherd* magazine, told me the details of this emergency treatment in a letter in 1966. They had used the procedure only a few times then, and cautiously. I have used it only when I felt that without it the lamb would surely die, and have found it always successful, but frightening to do.

PNEUMONIA IN LAMBS

Pneumonia is probably responsible for more lamb deaths than any other single cause (except starvation). In some flocks, it is responsible for as much as 12 to 15 percent death loss in the lamb population. For the most part pneumonia can be prevented. It is caused by drafts in cold damp quarters, by overheating with heat lamps followed by exposure to cold, and exposure to infectious agents from the ewe.

Proper management is the key to success in the prevention of pneumonia. Adequate ventilation in the lambing barn is mandatory. Open-sided barns with burlap bags or the new Tensar windbreak material to prevent drafts will stop a buildup of ammonia-laden stagnant air. Use heat lamps no more than is really necessary, and lambing pens with solid bottoms to prevent floor drafts on the newborn.

If pneumonia is a recurring problem in your young lambs, make sure that your selenium and vitamin E levels in the ewes are normal, as marginal levels result in immunosuppression and increased susceptibility to infection. Another treatment that has had remarkable success is the use of an intranasal vaccine, Nasalgen-IP. Parainfluenza III (PI-3) is a common viral disease of cattle that has been documented as a major cause of respiratory disease in sheep. While it is a highly infectious virus, infections can often be mild or even unapparent. However, under conditions of stress coupled with a bacterial exposure, it can trigger a high incidence of fatal pneumonia in both lambs and adults. The Nasalgen-IP vaccine is easy to administer because it is simply sprayed into the nostril. It functions on the same immunological parameters as the oral polio vaccine in humans. Each ewe and ram should receive 1 ml of the vaccine in a single nostril, and the lambs should be vaccinated during the first ten to eighteen hours of life, if possible, with the same amount. Although the vaccine has been shown to be safe, effective, and very beneficial, it has not yet been licensed for sheep. Your veterinarian can prescribe it for you.

MECHANICAL PNEUMONIA

Mechanical or "foreign body" pneumonia results when fluids or objects enter the lungs, such as excessive birth fluids or milk in the lungs of lambs. An abnormal birth position or any interruption of the umbilical blood supply to the lamb results in an oxygen deficit. This in turn stimulates the respiratory reflex, which causes the lamb to attempt to breathe before birth is complete. This causes inhalation of excessive volumes of fetal fluids, resulting in mechanical pneumonia. Also, forced bottle feeding of a lamb with impaired sucking reflex, improper stomach tubing, or oral medication can allow fluid to enter the lungs. There is no known cure for mechanical pneumonia.

SCOURS (DIARRHEA) IN NURSING LAMBS

Diarrhea in newborn lambs (called "scours") can be very serious, and has many causes. The yellow kind of scours is the least serious, and is usually caused from too much milk consumption: either from overfeeding by bottle, or because a strong lamb is sucking a mother who has an excess of milk. If bottle feeding, substitute a feeding with water or oral electrolyte such as Gatorade and/or reduce the amount of milk-replacer powder used in the mix until the condition resolves. If the lamb is nursing, milk out the ewe to reduce the amount of milk available, and give the lamb a feeding of water or Gatorade so that its appetite is satisfied for a feeding. A couple of teaspoons of Pepto Bismol will help firm up the droppings and form a protective coating in the intestine. Kaopectate is also helpful as an intestinal protective. If the scours continue for more than a day, an infection may be developing, and the lamb will need preventative treatment for dehydration and/or infection. The lamb should be given oral electrolyte solution, to replace the excessive electrolyte loss, and some form of antibacterial therapy. The tetracycline (terramycin/oreomycin) or sulfonamide preparations for baby pig scours are excellent for use in lamb scours. Many are available in a handy pump dispenser that can be kept in the refrigerator.

White scours are very serious and usually indicate an infection by *E. coli* which can result in very rapid dehydration, toxemia, and death if not treated immediately. In most cases it is caused by filth, such as poor sanitation; contaminated bottle, nipples, milk, feeders, or lambing pen; or a lamb sucking on a dirty wool tag from an uncrotched ewe.

If white scours is a recurring problem in your baby lambs, consider the new Ovine Pili Shield (from Grand Laboratories). One dose of this bacterin produces antibodies that are passed through the ewe's colostrum to the lamb, greatly decreasing the incidence of white scours caused by *E. coli*.

Sulfa or tetracycline antibiotic preparations are helpful for most of these bacterial infections, combined with doses of Kaopectate to coat the intestine. Some scour medications contain vitamins in addition to antibiotics.

With bottle lambs, discontinue milk feeding at once. For one day, feed either limewater (see recipe, Chapter 10) or Gatorade or a similar oral electrolyte (see recipe below) at the rate of 2 ounces about two or three hours apart. The lamb must have the liquid to prevent dehydration. Bottle-lamb white scours usually results from overfeeding coupled with contaminated feeders, or from milk-replacer left too long at room temperature. A colostrum-deprived lamb is very susceptible to bacterial scours.

The following is a recipe for a homemade electrolyte solution:

> 1 quart water
> 2 ounces dextrose
> 1/2 teaspoon salt
> 1/4 teaspoon bicarbonate of soda

Give this for only one day, or part of the day, until the diarrhea ceases, then return to milk feeding, giving smaller quantities than before.

In addition to the bottle of substitute milk, the bacterial scour treatment should also be given. You may be able to get the lamb to take some of it in a bottle.

For simple lamb diarrhea (not bacterial scours) it is often helpful to give a few ounces of aloe vera juice to help the digestive system return to normal.

To prevent bacterial scours, some producers give each lamb 1 cc of benzathine penicillin at birth, subcutaneously. While drugs are definitely useful in scours if needed, good management and sanitation will prevent many problems.

NAVEL ILL

This is a term used to describe infections from any number of organisms that gain entrance to the lamb's body through the umbilical cord, shortly after birth. They develop into serious illnesses, usually within a few days.

By treating the umbilical site with strong iodine *as soon as possible* after birth, and seeing that the lamb nurses its mother within the first hour (because the colostrum contains antibodies for many of the germs in its environment), you can minimize the danger of navel ill. A second dousing with iodine twelve hours or so later is a good practice. Clean bedding in the lamb pen will lessen the chance of infection.

Healthy little lambs.

The acute form of navel ill causes a rise in temperature, no inclination to suck, and usually a thickening can be felt around the navel. Death follows quickly.

Tetanus is one of the serious diseases caused by a bacillus that can enter through the cord. Certain protection against tetanus is obtained by vaccinating the ewes in the last two months of pregnancy (two separate shots, following label directions) with Covexin-8, which protects the ewe and passes protection along to the lamb in the colostrum. This vaccine is for tetanus, enterotoxemia, and all colostridial diseases that can strike lambs.

Since navel ill can be caused by various bacteria, it takes a veterinary diagnosis to determine the specific cause and thus administer the proper antibiotics. Treatment can consist of IV antibiotics, scour boluses, a tube passed to relieve bloat, and interperitoneal glucose.

CONSTIPATION IN LAMBS

A constipated lamb usually stands rather humped up, looking uncomfortable, with no sign of droppings, or only a few very hard ones. Sometimes the lamb will grind its teeth, and if the condition continues, will go into convulsions and die unless medicated. Administer two tablespoons mineral oil or one tablespoon castor oil, for a very small lamb (under two weeks old). For one as old as two months, give 1/4 to 1/2 cup mineral oil (carefully). Often the dosage will have to be repeated.

"Pinning" is an external kind of stoppage that is fairly common in very young lambs, usually under a week old. The feces collect and dry into a mass under the tail, gluing it down and plugging up the lamb. If not noticed and corrected, the lamb will die. Clean off the mass with a damp rag or a paper towel, trimming off some of the wool if necessary; disinfect the area if it is irritated, and oil it lightly to prevent further sticking. Check the lamb frequently. This is another good reason to keep mother and lamb in the pen for the first three days, so you can easily inspect for this and other problems.

Occasionally a lamb can suffer from a rare birth defect in which it is born without an anus. These lambs will often go undetected for the first few days, until the distended abdomen and discomfort is observed. Quick detection and surgery is the only treatment. If the birth defect is not too severe, the lamb can be saved.

ENTROPION
(Inverted eyelids)

Frequently, when a lamb is born, the lower (or sometimes the upper) eyelid, or both, may be rolled inward. When this happens, the eyelashes chafe the eyeball, causing the eye to water constantly, inviting infection and even blindness. The tendency is hereditary, but more prevalent in wooly-faced breeds. Do not keep such a lamb for breeding. Mark it with an ear tag or notch, for slaughter.

Inspect each lamb at birth, so that the condition is found at once and corrected. There are several ways to fix the condition. You can roll the eyelid(s) outward and hold in proper position by a clip, or sewing. The use of two little metal "surgical

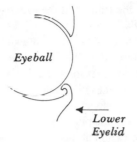

Entropion, showing lower eyelid turned in. Eyelashes will irritate the eyeball. Pocket behind eyelid becomes infected if condition is not remedied.

clips" is easier than stitching. They can be clipped into place with forceps or small pliers and left on for a few days.

If you want to try sewing them down, use white cotton thread and a sharp needle. Roll the eyelid out, put the needle through a small piece of skin and sew it down (the upper eyelid would be sewn to the forehead, and the lower eyelid to the jaw). In a few days the eyelids will have conformed to a normal position and stitches can be removed. Use a mild antiseptic in stitching or applying clips.

Another way of correcting the eyelid is with sharp scissors. With someone holding the lamb, grasp the skin below the eye with your index finger and thumb. As you do, you can see the skin unfold out of the eye. Remove this excess skin by snipping it with scissors. This leaves a small open sore that heals in a few days.

If you don't want to stitch or snip and don't have clips, pieces of Scotch or adhesive tape may suffice. They will have to be reapplied several times, to hold the eyelid in place for a few days until the condition is remedied.

Another treatment which I have heard about but not tried, is an injection of 1/2 ml of mineral oil (or preferably penicillin) subcutaneously into the lower eyelid with a tuberculin syringe and a 3/8-inch by 26-gauge needle. This creates a "bubble" between the conjunctiva (mucous membrane lining the eyelid) and the skin, which stretches the eyelid away from the eye. My veterinarian now uses this procedure (with common Pen-Strep penicillin) exclusively and has found it to be far superior to all previous surgical procedures. It takes him less than five seconds to correct the entropion once the lamb has been restrained by an assistant. DO NOT attempt this procedure without an assistant to hold the lamb's head perfectly still, and preferably only after being instructed by your veterinarian or an experienced shepherd.

URINARY CALCULI
(Stones, "Water Belly")

A problem of growing ram lambs over one month old, castrated or not, is that the salts they normally excrete in the urine can form stones, and these may lodge in the kidney, bladder, or urethra.

Symptoms. The lamb kicks at his stomach, stands with back arched, switches tail, strains to urinate, or dribbles urine, frequently bloody.

Some may recover if the stone is passed soon enough. This blockage of the urinary tract causes pain, colic, and eventually the rupture of the urinary system into the body cavity (hence the name "water belly"), and death.

If you are watching a lamb that appears to be straining and unable to urinate, put him on a dry floor for a couple of hours. Ordinarily he will urinate in that time, unless there is a blockage. Turn the lamb up and feel for a small stone that can be worked gently down the urinary passage. Sometimes manipulation of a small catheter tube (from the drugstore) will dislodge the stone.

Veterinarians say that nine times out of ten, the plugging is at the outer end of the urinary passage, so if the stone can be felt right at the end, and cannot be dislodged by gentle pressure, this outer end can be clipped, then disinfected. If the passage is cleared, and urine spurts out, stop the flow two or three times. It is possible for the bladder to rupture when it is emptied too quickly. If unable to dislodge the stone, a veterinarian may administer a drug that has a dilating action, or a smooth-muscle relaxer, to permit the calculi to pass, or may even remove the stone surgically.

Causes can be one or more of the following:

1. Low water intake due to cold weather or unpalatable water. Lambs need fresh warm water during cold weather. Help by adding salt to the ration, keeping both salt and fresh water in the creep. Increasing salt will increase urine volume and decrease incidence of stones.
2. Ration incorrectly high in phosphorus and potassium—like beet pulp, wheat bran, and corn fodder—and low in vitamin A. Correct by adding ground limestone or dicalcium phosphate, 1 or 2 percent of the ration, to make the calcium/phosphate ratio approximately 2:1. Well formulated lamb ration pellets have this ratio.
3. Growing crops under heavy fertilizer, with high nitrate content, interferes with the carotene in roughage that produces vitamin A. Vitamin A enrichment of ration would counteract this, and some of the lamb creep feeds have this.
4. "Hard" water may be partly the cause. Correct by adding ammonium chloride to feed, approximately 1/5 ounce per head per day, using technical grade. It is a harmless salt, and some pelleted feeds contain it.
5. Calculi are usually more prevalent when *only* pellets are fed, and seldom develop in lambs who get a level of 20 percent alfalfa.
6. Hormonal changes that occur when ram lambs are castrated at less than four weeks of age. The absence of testosterone after castration keeps the urethra from growing to its maximum diameter. If you have a persistent problem with your wether lambs, try castrating after six weeks.
7. Sorghum-based rations add to the risk of calculi. Cotton seedmeal and milo also increase risk. Corn and soybean meal are less apt to cause problems.

WHITE MUSCLE DISEASE
("Stiff Lamb")

White muscle disease in lambs is caused by insufficient selenium in the soil, and thus in the feed of the ewe, combined with a deficiency of vitamin E. When the soil is deficient (as in parts of Montana, Oregon, Michigan, New York, and many other areas) then the hay is also deficient in this important mineral. Hay from known localities with inadequate selenium should not be fed to ewes after the third month of pregnancy or during lactation, unless well supplemented by whole-grain wheat and mineralized salt with selenium in it. Treatment should also include vitamin E.

In areas that are known to be low in selenium, medication should be given to prevent lamb losses. BO-SE is an injectable containing both selenium and vitamin E, and is given to the ewe from two to four weeks before lambing, such as when you give the second Covexin-8 injection, which immunizes against tetanus and enterotoxemia.

L-Se is the same kind of medication, given to newborn or week-old lambs. These medications are made in two different strengths, so directions on dosage must be followed very closely, as too much selenium is deadly. Waiting to inject the lambs at birth instead of injecting the ewes will increase the number of lambs that could develop white muscle disease.

Symptoms. Lambs have difficulty getting up or walking, as they gradually become affected by muscle paralysis. If treatment is given soon enough, lambs will respond to L-Se, but once the muscle changes occur, they cannot be reversed.

Many small weak lambs, or lambs with a stiff neck at birth, will respond to treatment with L-Se.

ENTEROTOXEMIA
("Overeating Disease")

This disease is caused by a multiplying of bacteria called *Clostridium perfringens*, and can strike your biggest and best lambs, those who eat best. It is more common among lambs who are exposed to too much grain and too little roughage (hay), or who have had an abrupt change in their feed ration. It can also occur among fairly young lambs who are getting too much milk from a heavy milking ewe. The early creep feeding of both hay and grain will make these lambs not as prone to load up on milk. They may develop this disease if grain composes more than 60 percent of their ration, or if they are brought up to a full feed of 1 1/2 to 2 pounds of grain per day, too rapidly. It would appear that older lambs carrying a heavy load of tapeworms are especially vulnerable. Wet bedding, chilling, or stress can cause a variable feed intake that is conducive to disease outbreak.

Prevention is the very best plan. Vaccinate the ewe with Covexin-8, a 5-ml "priming dose" being given to her any time between breeding and six to eight weeks prior to lambing. The "booster" dose of 2 ml should be given approximately two weeks before lambing. In following years, the ewe will need only the "booster" dose. The colostral antibodies are passed to the lamb, providing immediate and complete protection against all clostridial diseases, including enterotoxemia and tetanus, the most common ones.

The immunity provided by the ewe, assuming that the lamb gets its normal amount of her colostrum, will protect the lamb until about ten weeks of age, when it should recline its own vaccination. Ideally, a "priming dose" of 5 ml would be given at nine or ten weeks of age, and a 2-ml "booster" about a month later. If this lamb is kept as breeding stock, it will need only the 2-ml "booster" dose about two weeks prior to lambing.

Enterotoxemia, if you have failed to vaccinate, is characterized by sudden death or convulsions and diarrhea. Few lambs live long enough to respond, but they could be treated with antiserum by injection. Prevention is the only sure way.

Some new pelleted lamb feeds contain chloro-tetracycline to control enterotoxemia, but immunization seems a healthier way to go.

PARASITES

Mature ewes eliminate millions of parasite eggs in their feces each day, so lambs are subject to infestation with parasite larvae from the pasture. While ewes should be wormed before lambing, the parasite problem does recur. The parasites seriously arrest the lamb growth, and a severe level of infestation can cause anemia and death.

One way of counteracting the situation is to provide a creep opening for the lambs into an adjacent "clean" pasture, with their own grain-ration feeder there, plus fresh clean water. Lambs should be well accustomed to grain and hay supplementation, and pasture, by four weeks of age.

It is standard practice to worm the lambs when they are separated from the ewes at weaning. When keeping sheep on overstocked pastures, the lambs may need worming once midway before weaning. Use Ivomec or Levamisole, or one of the other wormers that are safe for lambs, and note withdrawal time.

TETANUS
("Lockjaw")

Tail docking and castration can put lambs in danger of tetanus. If you have not boostered the ewes with Covexin-8, you should administer 300 to 500 units of tetanus antitoxin at the time of docking and/or castration. The antitoxin will protect the lambs for about two weeks, while the wounds are healing. Since there is no known cure for tetanus, protection is really worth the effort.

COCCIDIOSIS

Coccidiosis is an acute contagious parasitic disease which is spread between sheep by fecal contamination of feed or water. Strict sanitation and the arrangement of feed and water containers so as to prevent contamination will lessen the problem. Lambs must be prevented from walking in feeders and tracking manure into them.

Coccidiosis in lambs causes severe diarrhea, sometimes bloody, usually dark. Treatment would be medication added to feed or water. The disadvantage of adding medicine to water is in making it unpalatable, so that lambs would drink less water, and this could increase the danger of urinary calculi in the rams. Your veterinarian can check fecals for the presence of the coccidia oocysts (eggs) and prescribe Sulfamethiazine or Amprolium or any newer medications that may be developed.

Research at present indicates that the ionophore drugs are the best treatment, and it requires from seventeen to twenty-one days of continuous feeding before full effectiveness is obtained. Carrier animals cease shedding of infective oocysts so further spread is prevented. Lambs are usually kept on feed containing this drug until market weight, to prevent infections.

Many pelleted lamb feeds contain Deccox or Bovatec (lasolocid) and are fed continually to control coccidiosis and improve feed efficiency.

Bovatec is readily available, approved for sheep, and has a good safety margin. Deccox has proven effective, and can be mixed into salt at the rate of 2 pounds Deccox in 50 pounds of loose salt, fed free choice. Ewes should receive this continuously from thirty days before lambing until after the lambs are weaned. Rumensin, a cattle drug, is sometimes used, but it is not approved, being highly toxic in sheep if even the slightest mixing ratio error is made.

ACIDOSIS
(Grain engorgement, acute indigestion, "founder")

This is caused by the excessive production of lactic acid in ruminants that have experienced a sudden engorgement of grain or other high-carbohydrate-content feeds, or in feedlot lambs fed high grain/low roughage diets. Engorged lambs or those fed excessive grain-to-roughage rations become toxic from the excessive lactic acid that is produced by the fermentation of the high energy diet. As a result, the acidity of the rumen increases until severe digestive upset and/or death occurs. The sudden deaths of numerous lambs can mimic those caused by enterotoxemia, making accurate diagnosis difficult. Signs include inappetence, depression, lameness, coma, and death. At least 50-percent roughage (hay and/or pasture) is a safe ratio for the lambs, and any shift to a higher percentage of grain should be made very gradually.

POLIO
(Polioencephalomalacia)

Polio is a noninfectious disease of sheep, characterized clinically by blindness, depression, incoordination, coma, and death. The syndrome (which is similar to acidosis) is diet-related, but the exact predisposing mechanisms are unclear. It has been shown that the disease is caused by an acute thiamine (vitamin B_1) deficiency, and that the ruminal contents contain high levels of thiaminase (an enzyme that destroys thiamine).

British researchers, in 1974, found that all strains of a common rumen bacteria (*Clostridium sporogenes*) produced thiaminase, and this was the only thiaminase-producing bacteria recovered from material associated with field cases of polio. It is theorized that polio results when, by circumstance, the mix of nutrients in the rumen favors the growth of this organism and subsequent production of thiaminase in quantities sufficient to result in an acute vitamin B_1 deficiency. Empirically, field experience has shown that changing the ingredients in the diet may "upset" this particular balance, resulting in an alleviation of the outbreak.

Treatment with 0.5 gram thiamine hydrochloride will stimulate a rapid recovery if caught soon enough. Repeat treatments at two-day intervals as necessary. A lamb that has recovered can contract polio again, if its diet remains the same as before.

CHAPTER 12

Problems of Pregnant Ewes

PREGNANCY DISEASE
(Pregnancy Toxemia, Ketosis)

PREGNANCY DISEASE is highly fatal if not treated, or if the ewe does not lamb right away. When it occurs, it usually is in the last week or so of pregnancy, and often to twin- or triplet-carrying ewes. It can be reliably diagnosed by urine tests for ketones and acetones, with test strips from the drugstore or the veterinarian, but can usually be recognized without this.

It is possible to avoid this disease by using the ketone test strips early, to determine which ewes are not obtaining sufficient nutrition for their needs (see Ketone Test in "Feeds and Feeding" chapter).

Symptoms. Watch for sleepy-looking, dopey-acting, dull-eyed ewes, weak in the legs, with sweet acetoniec-smelling breath (it smells like model-airplane glue!). They will probably refuse to eat, then become unable to rise, and will grind their teeth and breathe rapidly. If treatment is delayed too long, recovery is doubtful.

Treatment. Four ounces of propylene glycol (this is not the antifreeze kind of glycol, which is poison), or four ounces of glycerine diluted with warm water, or a commercial preparation for treating ketosis should be given by mouth twice a day. Continue treatment for four days, even though the ewe appears to have recovered, to prevent relapse.

Along with the glycol treatment, some veterinarians say that an 8-cc injection of cortisone will shorten the recovery time, but this may cause premature lambing.

Keep propylene glycol (or the commercial medication) on hand before lambing, for prompt treatment of any suspected cases. Once a full-blown case occurs, it may have gone so far that treatment is ineffective and a cesarean section will be required to save the ewe. Loss of lambs will occur unless the ewe is very close to normal lambing time. Subclinical pregnancy toxemia (also confirmed by ketosis test) is the same disease, but it is a mild case where the ewe can be weakened and produce a small lamb or a dead lamb.

Cause. This is somewhat of a "vicious circle" disease, in that multiple or large fetuses require large amounts of nutrients, but as they grow and take up more abdominal room, the ewe's ability to consume sufficient feed to support both her and the lamb(s) is drastically reduced. The disease is not a "thin ewe" or blood sugar problem, but one of insufficient energy (calorie) intake. (My veterinarian refers to it as a "hypo-groceryosis" disease.) When the ewe is taking in less energy than is required to sustain herself and the lambs, she begins to use stored body fat to provide this energy. Ketones are the by-product of fat metabolism. If the ewe is breaking down significant levels of body fat, she may reach the point that the ketones are being produced faster than her body can excrete them. When this occurs, they build up to toxic levels and ketosis or "pregnancy toxemia" occurs. Ketosis often occurs in ewes that are on a high-fiber (hay) but low-energy (no grain) diet. Stress and/or forced activity also demands energy which can contribute to the problem or actually trigger the toxemia in borderline cases of inadequate nutrition. Simply stated, prevention requires *calories*.

TO PREVENT PREGNANCY TOXEMIA

- Avoid overfatness early in pregnancy.
- Encourage ewes to exercise daily.
- Provide rising level of nutrition in last four to five weeks of pregnancy.
- Supply a constant source of water.
- Feed regular amounts at regular times.
- Give molasses in drinking water.
- Avoid purchasing ewes too close to lambing.
- Avoid stress, and hurried driving of pregnant sheep.
- Make no sudden change in type of grain offered.
- Give special attention to nutrition of old sheep with poor teeth, late in pregnancy.
- Treat the feet of any lame ewe, or she may not move around well.
- Give nutritional grain combination, such as wheat/corn/oats, at least 1 pound per ewe, per day.
- Add molasses to the feed of all the ewes, if you have even one case of ketosis.

KETOSIS OR CALCIUM DEFICIENCY?

It can be difficult to tell the difference between pregnancy toxemia and hypocalcemia (milk fever). Pregnancy toxemia can be accurately diagnosed by test strips from the veterinary supply (or ketone sensitive strips from the drugstore). But this disease can be a complicating factor in a case of milk fever, so a diagnosis does not rule out calcium deficiency. You can make an intelligent guess by reviewing the circumstances:

If it is before lambing, and there is any possibility that the ewe may not have been fed properly in the last month, it is probably toxemia.

If it is after lambing, and the ewe is providing good milk for twins or triplets, and has had adequate feed with molasses, it is more likely to be primarily milk fever, but could have a trace of pregnancy toxemia as a complication. Most mail-order veterinary supplies stock commercial preparations for milk fever which also contain dextrose or other ingredients for ketosis, so they can be used to treat either or both.

MILK FEVER
(Lambing Sickness, Hypocalcemia, Calcium Deficiency)

Because so much calcium is needed to form the bones and teeth of the lambs, and so much calcium goes into the ewe's udder of milk, she suddenly may be unable to supply it all, due either to simple deficiency or deficiency triggered by metabolic disturbance, and this deficiency can cause death in a short time. The ewe's lack of sufficient calcium happens more often *after* lambing, but can be just before. It may be brought on by an abrupt change of feed, a period without feed, or a sudden drastic change in the weather.

Symptoms. The onset of the disease is sudden, and progresses rapidly. Earliest signs are excitability, muscle tremors, and stilted gait, which are followed by staggering, breathing fast, staring eyes, and dullness. The ewe next lies down and is unable to get up, then slips into a coma followed by death. Although this is called a "fever," the temperature is normal or subnormal, and the ears become very cold. To be successful, treatment should start before the ewe is down, but even when she is down, prior to coma, there is a chance of recovery.

Treatment. Milk fever is most often a true "medical emergency" in which life or death of the animal is a race against time. Once the condition is sufficiently advanced, *intravenous* treatment is the only route of injection that will save the animal. If in doubt, call your veterinarian immediately.

If you cannot give an intravenous injection and/or veterinary assistance is not available, the ewe should be injected subcutaneously with 75-100 cc (divided in five places) of calcium borogluconate or calcium gluconate. The latter is cheaper, easier to find, and equally effective. You can purchase it at a drugstore. Tell them it is for livestock use.

The injection should be intravenous, but while sub-cu gives a slower reaction, there is less chance of cardiac arrest (heart failure), and it is a safer procedure at home. It can be given intraperitoneally, in the paunch on the right side of the ewe. Lay the ewe on her left side, and inject into the right side if she has already lambed. This is faster than sub-cu and safer than IV.

Several commercial veterinary preparations for treatment are sold in sheep-supply mail-order catalogs.

If milk fever comes on before lambing, it can be confused with pregnancy toxemia. If it is a calcium shortage, however, the ewe will show a dramatic improvement after calcium is given. I saw this only once, with a mother of triplets, and her recovery was very fast, after treatment.

VAGINAL PROLAPSE

Prolapse of the vagina is most common before lambing, but can occasionally follow a difficult labor. The vaginal lining will be seen as a red mass, protruding from the genital opening. Do not delay treatment, for it will get progressively worse and more difficult to repair. Early detection is important. Even though this problem occurs infrequently, be on the watch for it.

Treatment. If the lining is just barely protruding, confine the ewe in a crate that elevates her hind end, thus decreasing the pressure. Leave her head out to eat, and feed her mostly on grain, plus some green feed (grass, weeds, apples, etc.). Avoid ground-up grain or rolled oats. The dust might cause coughing and aggravate her problem.

A more certain solution, still best begun as early as possible to be successful, is to use a *prolapse loop*, or *prolapse retainer*, which is a flat plastic tongue, often called a *ewe bearing retainer* (available from livestock-supply catalogs, see Sources).

If you have already ordered your supplies ahead of lambing season, you are ready to treat the problem if necessary. If not, here is an alternative; the late Dr. Beck gave the above design for a homemade prolapse loop, easily fashioned from wire. The loop is made from about 1/16-inch wire with only the loop part covered with soft rubber tubing as a cushion, the rubber tubing being slipped onto the wire before making the final bends. Bend as shown in the drawing, and disinfect the whole loop before use.

Before attempting to insert the prolapse-retaining loop, wash the ewe's prolapse, and place her with her hind end considerably elevated. You can tie the ends of a length of rope to each hind leg, and loop the rope up over the top of a post, or have the rope short enough to just go around your neck, so that the ewe is raised by it. You can hold her steady in this position if she is on her back, with the hind end raised and steadied by a bale of hay.

To replace the vaginal protrusion and insert the loop or retainer:

1. Cinch a rope or belt around her middle so that she cannot strain after you replace the prolapse. A 1/4-inch rope around her flank in front of the udder, not so tight that she can't lie down and get up, will do. This will have to be removed when she goes into labor. Sometimes she will stop straining after a couple of days, as the swelling goes down.
2. Wash your hands, and disinfect the loop if you have not done this already.
3. Wash the prolapsed part with cold (not hot) antiseptic water, or put both mild soap and antiseptic in the water to help contract it.
4. Watch out for a flood of urine as you gently replace the vaginal lining. Its bulging may have blocked the opening to the urinary tract. This would cause death if blocked too long.
5. Replace the lining, using lubricant if necessary, and press out all the creases gently. This is much easier with the hind-end-elevated position than it would be if she were lying flat. Even holding her on her back, with her shoulders on the ground and her hindquarters up against your knee, will relieve much of the pressure against the replacement.

Materials: 1/8"-3/16" aluminum wire or rod; rubber tubing to fit over loop.

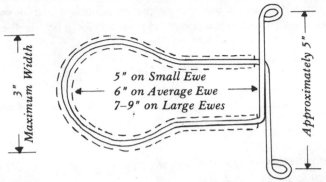

5" on Small Ewe
6" on Average Ewe
7-9" on Large Ewes

Maximum Width 3"

Approximately 5"

Homemade prolapse loop. Wire "loop" is covered with soft rubber tubing as a cushion. Put on rubber tubing before making last bends in the wire. Dotted lines show where rubber tubing is placed over wire.

6. Holding the vagina in place with one hand, insert the prolapse loop, straight in, flat horizontally. If you have made a loop from the pattern given, it should be long enough that the forward end is in contact with the cervix at the end of the vagina.
7. The loop is held in place by tying it to clumps of wool, or by sutures if the ewe has been sheared or closely crotched. There is also a new Prolapse-Harness (see Sources) which can be used with the prolapse loop, holding it in place better than tying to the fleece.
8. Give a shot of antibiotic to avoid infection.

The ewe can lamb while wearing the loop or retainer, or you can remove it as she goes into labor. It is safer to leave it in place and try lambing that way, so that the ejection of the lamb does not start her problem all over again.

Mark this ewe for culling, for the prolapse produces permanent damage and might happen again. Since it could be a genetic weakness, it is best not to keep any of her lambs for breeding.

Sutures. At one time the standard handling of prolapse was with sutures to hold the vagina in. These must be removed at lambing time. We have successfully used dental floss and a curved needle, with pliers to get a good grip on the needle and pull it through.

The safest way to suture is to use only one deep stitch at the top of the vaginal opening, and one across the bottom. Insert the needle from right to left at the top, then bring the needle down and insert it from left to right at the bottom. Knot the two ends together on the right side. The advantage of sewing this way, rather than crossing the stitches across the center of the opening, is that you can tell when the lamb is coming. There is room for the feet and nose to present themselves, allowing you time to cut the stitching.

And if prolapse occurs after lambing, the replacement and retaining with sutures would be the most logical approach.

Induced labor. If you know the lamb due date and the ewe is at least 143 days along, your veterinarian can prescribe or give medication such as dexamethasone to start labor, after which replacement and suturing would be effective.

Prolapse Causes

- Anatomical weakness, likely inherited.
- Feeding too much roughage during late pregnancy, with lambs and stomach causing excess pressure, combined with weakness in that area.
- Dusty hay or crushed grain, making ewe cough.
- Deficiency of vitamin A.
- Extra-fat ewe, lying down in normal sheep position on an upward slope, facing upward, causing too much pressure.
- Pneumonia or lungworms, causing ewe to cough a lot.
- Overly short tail-docking, weakening the muscle structure.
- Rough handling in shearing or worming during late pregnancy.

Prolapse Prevention. In addition to the previously mentioned causes, another factor is now known to be important. The need for selenium is recognized and its use is not new, as it is known to increase lamb survival and prevent White Muscle Disease in lambs. Now it is seen to increase muscle tone and help counteract a prolapse tendency in pregnant ewes.

Experts at the USDA Sheep Station at Dubois suggest that anyone in a selenium-deficient area should be injecting ewes with selenium one week before lambing, for muscle tone. Because of the potential trauma to ewes, injection could be earlier, giving the selenium-E at the time you give the last Covexin-8 vaccination. The slightest indication of the beginning of an actual prolapse would call for an additional injection of selenium, along with the usual prolapse-repair measures. While selenium-vitamin E injectables have instructions calling for intramuscular injection, many experienced sheep veterinarians recommend subcutaneous injection to avoid incidence of muscle damage at the injection site. Covexin-8 is also given sub-cu, but a vitamin ADE shot must be intramuscular because of its adjuvant base.

A ewe who has had a selenium-supplemented ration or selenium-E injection will usually have an adequate level of plasma selenium and will produce milk with sufficient selenium. Too much selenium is acutely toxic, so you would not want to have a selenium-enriched ration, plus mineral-salt mix with selenium, plus injections. There are several selenium products available, but all with different concentrations, such as LSE with .25mg per 1 ml, BOSE with 1 mg per 1 ml and MUSE with 5 mg per 1 ml. Check the concentration carefully before use, and follow label. A newborn lamb, for instance, should never be injected with more than 1.0 mg of selenium.

There is reason to anticipate that areas of selenium deficiency will be getting worse, and new shortages appearing. Increased forage yields are speeding the depletion of selenium in the topsoil, and increased animal stocking per acre on a given land area is also contributing to the problem.

Anyone wondering about their particular area should check with the county extension agent. Blood tests can give an accurate check, so that proper supplementation can be done on an informed basis.

ABORTIONS

These can be caused by circumstances. Moldy feed, with mold spores infecting and destroying the placenta, can cut off nourishment to the fetus.

Injury is often a cause, such as when a ram is running with the pregnant ewes and bumps them away from hay or feed. Narrow doorways, where sheep rush through for feed, are also dangerous as the ewes become large with lambs. Dog attacks nearly always cause abortions among the ewes who have been injured or chased.

When a ewe has lost her lamb through abortion in the last few weeks of pregnancy, or has a stillbirth and there is no orphan to graft on her, she should be milked out on the third day and again in a week, if she has a full udder. If the lamb was born dead due to a difficult birth rather than disease, the first milking should be done at once and the colostrum frozen for future use. With a tame and docile ewe, you may want to continue the milkings for a while and then taper off, saving all the milk for future bummer lambs.

The other causes of abortion are disease, and these causative diseases are being more defined now than in the past. There is now an EAE-Vibrio combination vaccine which can be administered two weeks before breeding, with a booster shot each year two weeks before breeding. This can protect against the two following diseases that often cause abortion:

Vibriosis is caused by bacteria that can live in the gall bladder and intestine of the ewe, but invade the uterus, placenta, and fetus during late pregnancy. Although ewes that have aborted from this are immune to further abortions, they can be carriers that contaminate the feed and water, infecting other ewes. The rest of the ewes can be vaccinated, followed by three daily injections of 8-cc pen-strep, followed by 500 mg a head per day of chlortetracycline (CTC) until the lambing period is over.

Enzooatic abortion of ewes (EAE) is caused by an organism called *Chlamydia* which causes late term abortions, stillbirths, and weak lambs. It is not the same species of *Chlamydia* that causes respiratory diseases, epididymitis in rams, conjuctivitis (pink eye), or arthritis in sheep. It is an obligate parasite (does not live freely in the environment) of sheep, spread to susceptible ewes by contact with aborting ewes, infected fetal membranes, uterine discharges, or a dead fetus. Susceptible ewes thus infected will most likely abort the following year unless infected early in their gestation, in which case it would happen that year. The only treatment that is currently advocated for an EAE outbreak is the feeding of 500 mg CTC per ewe, daily. Vaccination after an outbreak occurs is not helpful.

Since symptoms of vibriosis and EAE are similar, it takes analysis from a Livestock Disease Diagnostic Laboratory to identify the exact cause of an abortion. The EAE-Vibrio vaccine protects from both.

Toxoplasmosis is a microscopic protozoan (coccidium) of cats. In unnatural hosts such as sheep and other species (including humans), the organism "gets lost" in its normal migratory route and invades many tissues causing infections in the brain, eyes, uterus, fetal membranes, and the fetus itself. Clinical signs are consistent with the damage caused to the particular tissue. Abortions and stillbirths are most commonly observed. Infection occurs when cats defecate, leaving the infectious organism on hay, grain, and other food consumed by sheep. Ground-up grain is a common target because it constitutes a ready-made "sand pile." There is no vaccine or effective treatment. Once diagnosed on a farm, the only preventative is to rid the farm of cats. Stray cats should not be allowed to stay. Strict sanitation, clean uncontaminated water, dry protected storage of hay and grain, and off-the-ground feeding troughs will reduce the incidence and spread of disease.

RETAINED AFTERBIRTH
(Retained Placenta)

In almost all cases, the afterbirth comes out normally, usually within the first hour after the lamb is born, depending somewhat on the activity of the ewe. Do not try to pull it out, if it is hanging part way out, as you might cause the ewe to strain and prolapse, or to do other injury to herself. You can allow quite some time to pass, without worrying about the afterbirth, for a veterinarian does not consider it a real "retained placenta" until six hours have passed since the birth of a lamb. Some ewes eat the afterbirth if you are not there to remove it, causing you to think it is retained.

If more than six hours have passed, home treatment can consist of an injection of streptomycin or penicillin to ward off infection. Forcible removal of the afterbirth is best done by the veterinarian who can differentiate between the maternal and the fetal cotyledons, to separate them. It is usually not advisable to remove the placenta manually sooner than forty-eight hours after the birth, and the veterinarian may in the meantime give a drug that could assist in expelling it.

Causes

- Exhaustion following difficult lambing.
- Nutritional disorder, such as deficiency of selenium, magnesium, or calcium, which can affect the ability of the uterine muscles to contract properly.
- Premature birth, sometimes result of poor feeding in the last four weeks of pregnancy.
- Infection or abortion.
- Hereditary weakness.

MASTITIS

This is an infection and inflammation of the udder, usually affecting only one side, and can be caused by one of several different bacteria.

In active cases, the ewe has a high fever (105 to 107 degrees) and usually goes off her feed. One side of her udder is hot, swollen, and painful. She will limp, carrying one hind leg as far from the udder as possible, and not want her lamb to nurse.

The milk becomes thick and flaky, or full of curds, or watery. Early detection and prompt treatment can minimize udder scarring.

One type of mastitis results in gangrene. The udder is almost blue and is cold to the touch. Large and repeated doses of Digydrostreptomycin may be helpful, but this type of mastitis is critical and an IV is needed in order to save the ewe, who should be marked for culling.

Another type of mastitis will respond if penicillin treatment is given early enough, usually in dosage of 500,000 to 1,000,000 units. Your veterinarian can examine with a microscope the exudate from the udder and know which organism is the cause, then prescribe proper treatment. If the ewe is treated aggressively at the first signs of mastitis, there is a 50-percent chance of saving the udder.

Milk or subclinical mastitis may be undetected, showing up at the ewe's next lambing when she has milk in only half of her udder, and the other half is hard. Milk cases are most often caused by bruises from large lambs almost weaning age. They bump their mothers with great zest as they nurse, sometimes lifting her hind end right off the ground, or twins who pummel her simultaneously. Mild mastitis has fewer symptoms, and the ewe may just wean the lambs by refusing to let them nurse.

Causes

- Undue exposure to cold and rainy weather, ewe lying on cold wet ground.
- Infection from an active case of mastitis in another ewe—indiscriminate nursing by some lambs (not getting enough milk) may transfer bacteria or transmitted by flies having fed on infected teats.
- Udder injury from high thresholds in barns, or from underbrush.
- Udder injury from large nursing lambs.
- Large milking udder, reducing resistance of udder to bruises and infection.
- Loss of lamb, ewe with large milking udder, not milked out occasionally to dry up milk production.
- Sudden weaning of lambs while ewe still has full milking capacity.
- Grain not withdrawn at least five days prior to weaning.

Sheep people say that mastitis is governed by Murphy's Law: The severity and incidence of mastitis is directly proportional to: (1) the value of the ewe; (2) her lack of age; (3) the number of lambs she delivered; (4) the severity of the weather; and (5) how busy you are at lambing time.

Treatment. Antibiotics given by injection. Also, infected side milked out completely and milk destroyed, and antibiotics inserted into the teat. Veterinary supply catalogs sell the teat medication for cows, and while the applicator is a little large for convenience, it can be used with sheep. There are combination treatment drugs, for both acute and mild chronic mastitis, and these are effective against several of the causative bacteria.

The affected half of the udder is not likely to have milk again unless mastitis is noticed early and treated promptly.

CHAPTER 13

Internal Parasites

IN COMPARISON to other animals, sheep are more resistant to bacterial and viral diseases, but more susceptible to internal parasites. A weakened and run-down condition from parasite infestation can be a principal cause of a disease outbreak.

A heavy load of parasites is a vicious cycle leading to undernourishment of the sheep, which in turn makes them even more vulnerable to parasite damage. The highest death losses occur in lambs, yearlings, and extremely old sheep, with death loss higher in poorly fed sheep. Internal parasites ("worms") reduce productivity and cause anemia, wool break, progressive weakness, and death.

LIFE CYCLE OF WORMS

There are approximately ten species of worms that cause problems in sheep. For the most part, they all have a similar life cycle. The various species of worms live in the true stomach, small and large intestine where they feed on blood and body fluids from the lining of the digestive tract, causing anemia and serum loss. Millions of eggs from these worms pass out with the feces, and under favorable conditions of warm weather and moisture, hatch out into infective larvae in about five days. These larvae migrate onto the moist sections of the grass and are ingested by the sheep. Once swallowed, they invade the tissues of the digestive tract where they undergo a maturing stage and emerge as adult worms in about twenty-one days. Most of the eggs and/or larvae are killed under conditions of cold freezing temperatures or hot dry weather, which in many parts of the country "sterilize" the pasture. However, nature has devised a diabolical survival mechanism for these worms that allows them to overwinter or otherwise survive periods of adverse conditions by hibernating as immature worms in the tissues, only to emerge weeks or months later when conditions for survival are more favorable. More on these hibernating larvae later, as they figure strongly in the design of a more effective worming program.

IMMUNE RESISTANCE TO WORMS

Constant exposure to migrating worm larvae over time results in the development of a degree of resistance to worm burdens in older sheep because the larval proteins act as a form of "vaccination" against the larvae. Technically speaking, this "immunity" is actually the development of the antibodies that cause allergic reactions. In brief, a mini-allergic reaction occurs in the tissues surrounding the incysted worm larvae, in which a combination of smooth muscle contractions and fluids cause the parasites to be dislodged and expelled into the lumen of the intestine. They then pass out with the feces. This partial immunity to worms takes about two years to develop fully, which explains why older ewes do not accumulate as much worm burden as lambs on the same pasture, and why lambs must be wormed more often than the older ewe population.

PARASITE CONTROL

The main factor contributing to heavy parasite loads is population density. A few sheep on a given acreage will deposit less eggs, hence fewer infective larvae, than a large number of sheep on the same area. If you are able to rotate the sheep from one pasture to another, you can allow time for the worm larvae to die from age and exposure on the recently contaminated grass. The eggs/larvae of many stomach worms can easily survive three months in cool damp weather, but much less in dry hot weather. The old Scottish rule of thumb is, "Never let the church bell strike thrice on the same pasture." This means that if you live in a moderate climate, and have a choice of pastures, don't turn sheep out in the spring onto the last fall-grazed pasture. Select one that was last used in the heat of summer. Unless pasture rotation allows time for the death of the larval stages in the grass, rotation is of more benefit to the pasture growth than to worm control.

Overstocking of pastures cuts down the feed supply, which weakens the sheep. It also causes them to crop the grass more closely, ingesting more larvae to increase the worm load. It aids parasites in completing their life cycle.

Sheep in poor nutritional condition cannot tolerate as much worm burden as well-nourished sheep can. Lack of proper diet, insufficient protein, and incorrect balance of nutritional elements, including lack of vitamins and minerals (such as selenium) can leave them more vulnerable to worm damage.

One step toward better worm control is sanitation. Never put feed directly on the ground where it can become contaminated. Make sure that the water supply is clean and protected from fecal contamination. Sheep like fresh clean water. If it becomes fouled, they may not drink enough for maximum production or to supply their body needs. The lambs' creep feeder should have a baffle above the trough to prevent the lambs from climbing into the grain.

It is necessary to recognize symptoms of a worm build-up and carry out an adequate control program with the appropriate medications.

WORMING

With the development of safer and more effective worming drugs such as Tramisol, Thiabendazole, Fenbenzadole, and Ivomec, worming can be done more effectively and without harming pregnant ewes or small lambs.

Ewes should be wormed at the beginning of the flushing period, two or three weeks before breeding. They will not settle properly and will have a more protracted lambing period if they must contend with a load of parasites. Also, the ewes will produce fewer twins and more weak lambs, and will have less milk for the lambs. Pregnant ewes with worms are drained of needed energy, and their weakness leaves them more susceptible to pneumonia or pregnancy disease, and too weak to withstand a difficult delivery.

Even though the ewes may be essentially worm-free, going into the winter, and are housed in dry lot where they are free from reinfection, they will still experience a worm burden shortly after lambing. This postlambing rise in parasite load is due to hormonal changes which trigger the incysted larvae to "wake up" and complete their normal life cycle. A similar rise in worm burden also occurs in rams as spring approaches. This is the worm's survival mechanism that I spoke of earlier.

Worming the ewes at approximately three weeks after lambing reduces the burden of worms that are competing for the ewe's energy requirements for milk production. In areas where weather conditions are favorable for larval development from the eggs shed in the feces, this worming also reduces the hazard to the lambs, and aids in preventing the postlambing, worm explosion. It may be necessary to worm two or three weeks prior to lambing, in climates where worm infections can occur during gestation. Handle the ewes carefully, for the stress of catching and worming in the last few weeks of pregnancy could combine with some slight nutritional deficiency to trigger pregnancy toxemia in ewes carrying twins or triplets. Levamisol (Tramisol) or Ivomec is good for this late pregnancy worming, because they have some effect against hypobiotic (arrested) and migrating larvae. Other wormers kill only the adult forms that are in the intestine, allowing the migrating larvae to again begin "setting up housekeeping" the day following worming.

For the most part, worming medications have *no residual activity*. They are only good the day you give them. Following clearance of the medication from the intestine, the immature larvae are free to move in and begin building the worm burden all over again.

TARGETED WORMING

In most sheep-rearing areas, the worm population is severely depressed or completely killed off during the winter months. This then means that approximately 95 percent of the worm population is in the sheep in the form of hypobiotic (arrested) larvae incysted in the tissues, with less than 5 percent out on the pasture. Once the sheep gain access to the pasture during favorable weather conditions, the ratio reverses, with 95 percent of the worm population in the grass and 5 percent in the sheep. It is possible to worm the sheep, but impossible to worm the pasture. Logic then dictates that the most opportune time to deal a severe blow to the new season's worm population is to reduce it as much as possible in the sheep prior to the grazing season, so that the sheep cannot transfer or "seed" the population back to the pasture. The sheep should be wormed two or three days prior to turning out on pasture so that the worm eggs already in the feces have an opportunity to fall harmlessly in the barn or lot where the larvae cannot survive.

Many people practice a double-drenching in the summer, which they say reduces the worm burden for the season. This is not always the case in areas of constant rainfall or irrigated pastures where the temperature and moisture conditions can provide a consistent, high-level challenge throughout the summer. In areas where the grass begins to dry, worming the ewes at that time, followed up with a worming six weeks later, will normally reduce the worm burden below harmful levels. The hot dry weather significantly reduces the larvae population in the pasture, thereby reducing the infection rate in the ewe. It is also helpful to move sheep to a clean pasture twenty-four to forty-eight hours after worming, to help keep the pasture clean.

LAMBS AND PARASITES

Young lambs that are turned out to pasture with ewes pick up worms that will grow to maturity in about a month. As the worms increase they cause anemia and even death. The dying lambs may not be thin, as the effects of severe anemia come on very quickly in the younger animal. You can prevent anemia by worming the lambs at about two and a half to three months of age (note withdrawal days before slaughter, on label).

To prevent severe infestation, practice what is called "forward creep grazing." When rotating pastures, let the lambs graze each clean pasture ahead of the ewes, through a creep-type fence opening of a size that the ewes cannot get through.

PARASITE REDUCTION MIXTURE

While TM (trace mineralized) salt does improve the growth of the lambs and the health of the ewes, it is believed to also encourage the growth of the internal parasites. This can be counteracted by adding dicalcium phosphate to the loose mineral salt (fed free choice in a salt box). Since this is a good source of calcium for the lambs, it should be done anyway.

A popular mixture is six pounds TM salt, three pounds dicalcium phosphate, and one pound phenothiozine wormer granules. This low-level feeding of phenothiazine keeps worm eggs from developing properly. The resulting kill of eggs and larvae cuts down on the source of parasite infection on the pasture. No immediate results will be seen as it does not affect the eggs and larvae already on the pasture. However, over a period of time (combined with other measures) there will be fewer worm larvae on the pasture to reinfect the sheep.

SYMPTOMS OF PARASITES

One visible sign of worms is "bottle jaw" (swelling under the jaw). It is a final warning that the sheep have worms severe enough to cause deaths.

Other symptoms are diarrhea (for some kinds of worms) and anemia (for most kinds of worms). Anemia is indicated by the very pale color of the inner lower eyelids and gums, caused by intestinal worms drinking the sheep's blood, as much as a pint a day in heavy infestations.

There are eight or more kinds of small stomach worms (round worms) that cause anemia but not diarrhea. The sheep become listless, with pale mucous membranes,

"Bottle jaw" from severe stomach worm infestation. (USDA)

and lose condition, wasting away and dying if they are not wormed.

The small brownish stomach worm "ostertagia" causes scours. This worm is so perfectly camouflaged against the walls of the sheep's small intestine that it is difficult to spot in a postmortem.

BROAD-SPECTRUM WORMERS

Since most infestations are of more than one kind of worm, the broad-spectrum worming drugs are recommended for general worming. If necessary to use two different kinds of wormers for specific known problems, avoid alternating them too often lest you encourage wormer resistance at an accelerated pace. For most purposes, one of the several major broad-spectrum, low-toxicity wormers will take care of the most prevalent of the stomach parasites.

HOW TO ADMINISTER WORMING MEDICATIONS

Follow label directions for dosage and type of administration (as well as withdrawal days before slaughter). Wormer drugs are given as:

- Boluses—large pills.
- Drenches—fluids given by mouth.
- Granules or powders—added to salt or feed.

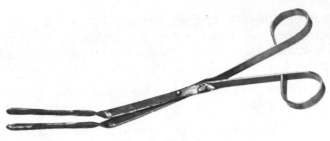

Capsule forceps.

- Premixes—already mixed into feed.
- Pastes—easy to use.
- Injections—follow label instructions.
- Blocks—usually with salt.

Boluses. For those without experience, boluses are easier and safer to give than drenches (fluid by mouth). Be sure you are buying the sheep-size bolus, for some wormers are also made in cow size. There are three methods of putting a bolus down a sheep:

1. By hand. Hold the mouth open far enough so that you don't get your finger crunched by the back teeth. With one finger, push the bolus until it slips back down the throat.
2. With a bolus gun (see Chapter 16 on its use, under "Oral Medications").
3. With capsule forceps, which make it easy to hold bolus. Straddle the animal, with her neck between your legs. Open the lower jaw with your left hand, using the thumb as a wedge to hold the mouth open. Be careful you don't injure the sheep with the forceps, or let the sheep bite your thumb. Deposit the bolus as far back as you can, withdraw the forceps, and hold the sheep's mouth closed for a minute. Don't release the sheep until you know the bolus has been swallowed.

Drenches. A ball drench gun, or automatic drench gun, is a liquid wormer, intended primarily for large numbers of animals. The person giving the drench carries a large bottle or canister of the medication strapped to his back, attached by a tube to the drench gun. Care should be taken not to penetrate the mucous membrane of the throat by overly forceful drenching. Depositing the medication behind these membranes may cause an abscess. The drench gun is efficient, but it is an initial expense warranted only by multiple use.

For a small flock, use a handy and inexpensive 2-ounce dose syringe, which is a small aluminum pipe attached to a rubber bulb. To administer the liquid drench, have the sheep in a standing position. *Keep its head level*, muzzle not raised above the eyes, or the drench will go into the lungs, causing death. Do not drench while the sheep is coughing or straining. Drenching should be gentle and slow.

Powders or granules. Phenothiazine, as a low-level parasite-control measure, is mixed with salt and dicalcium phosphate, and offered to the sheep free choice. This is done in addition to planned wormings with broad-spectrum drugs. The pheno-salt mixture should be protected either in the barn or by a roof shelter out in the pasture. Rain washes the mix away, and sunshine reduces its potency.

Two of the commonly used drugs for prevention of coccidiosis are formulated for mixing into feed.

Premixes. These are intended either for low-level use (as the lamb-pelleted feed in a creep feeder) or for use in a planned worming, as needed.

Paste. This would be sold with an applicator, so you smear it on the sheep's tongue.

Injection. Be sure to follow label directions as to the site for injection, the type of injection, and the dosage as given for sheep weight.

WORMER MEDICATIONS IN COMMON USE

These are not all (at this writing) approved for use in sheep. However, many are safe, effective, and commonly used (with veterinarian approval and prescription).

Tramisol® (levamisole). Effective against three species of stomach worms, six species of intestinal worms, all strains of the barberpole worm, and against lungworm, a type of worm not adequately controlled by many of the other worming drugs. At recommended dose, it is safe for pregnant ewes after the first thirty days of pregnancy, and for older lambs. This medication is sold as oblets (slightly smaller than boluses) or as a drench or injectable. Withdrawal time depends on form used. This wormer is approved for sheep. It causes paralysis of worm muscles.

Thiabendazole® (TBZ, OmnizoleR). Safe and effective, for stomach worms, large and small intestinal worms. Long-term usage usually results in some degree of wormer resistance. Approved for sheep use, with thirty-day withdrawal. Usually available as bolus, paste, drench, or feed additive.

Phenothiazine. Used as a drench or in boluses is no longer a standard treatment. Broad-spectrum wormers are used instead for the seasonal wormings. However, pheno is excellent for continuous low-level use, mixed into your salt formula. It renders infertile many of the worm eggs that are dropped onto the pasture.

Ivomec® (Ivermectin). Injectable or drench (note the withdrawal period for each). Some say injection not as safe as in a drench. Field trials conclude it was as effective at one half dose as at full dose level, but at one half dose the parasites could develop resistance faster. Effective against stomach worms, large and small intestinal worms, lung worms, blood-sucking lice, and keds (ticks). Has slow effect on nose bots, takes about thirty days to see results on them, using full dosage. Not effective against tapeworms.

Fenbenzadole (Panacur®, Safegard®). Safe and effective against stomach worms, large and small intestinal worms, lungworms, and tapeworms. Thirty-day withdrawal.

Equipar® Equine Wormer Suspension (Oxibendazole). Very similar to Panacur and Safegard in efficacy and safety. Does not kill tapeworms.

Telmin® (Mebendazone). Safe and effective for stomach worms, large and small intestinal worms. Some effect on tapeworms and liver flukes. Withdrawal thirty days. Not yet cleared for sheep, but your veterinarian may prescribe it.

Curatrem® (Clorsulon). Excellent treatment for liver flukes. Kills developing flukes as well as adult, but if sheep has much liver damage, would still not recover completely. Not yet cleared for sheep, but your veterinarian may prescribe it.

Rumatel® (moratel tartret). Safe and effective for stomach worms, large and small intestinal worms. At least thirty-day withdrawal time. As a pelleted dewormer, in tests it cleaned up 90 percent of ewe infestation and decreased egg count 97.5 percent. Adequate feeder space (12 to 15 feet per ewe) needed to be sure that all animals consume the wormer. Not yet cleared for sheep, but your veterinarian may prescribe it.

Bovatec® (lasalocid). New drug for prevention of coccidiosis, now approved for sheep use. Lasalocid, the active ingredient in Bovatec, works by killing the coccidia in early stages of developmemt so they cannot reproduce. No withdrawal time on medicated feed pellets.

Deccox® (decoquinate). Also to prevent coccidiosis, but not yet approved for sheep. Used in free-fed salt/mineral mix.

UNAPPROVED DRUGS

Unfortunately, we do not have as many options in "approved" drugs as do other countries such as Australia and New Zealand, which have larger sheep industries. Drug companies here know it is not practical to spend the time and fantastic amount of money involved to get approval of a medication for sheep, which are considered a "minor species." However, your veterinarian is allowed to prescribe any drug he can legally obtain, for "extra-label" use in any species, if a legitimate client-patient-veterinarian relationship exists. This is a crucial legal point. If a sheep producer uses an unapproved drug on his or her own, such use is not only illegal, but also the producer totally assumes legal and financial responsibilities if residues are produced and detected. Your veterinarian can prescribe any needed medications, and advise as to the proper dispensing, dose, and withdrawal times. Actually, many unapproved drugs, especially wormers, are in very common usage. There is much pressure on the government to make it easier for a drug which is already approved for one livestock species to become approved for another species. This would take much less investment by the manufacturers. The latest estimate of cost of research and development—involving chemists, microbiologists, toxicologists, physicians, veterinarians, and others, and taking as long as seven and a half years—averages up to $3.5 million per product, to get the drug approved for just one livestock species.

The most practical approach to parasite control is to use a wormer that has specific sheep-use labeling (approved). Follow label directions, warnings, and withdrawal time. By following up with after-treatment feces analysis by your veterinarian (worm eggs cannot be seen by the naked eye), it can be determined whether the product was effective. If not, your veterinarian can assist you with one of the off-label products.

LESS COMMON INTERNAL PARASITES

Lungworms. Prevalent in low-lying or wet pasture, and live in air passages, causing accelerated breathing, coughing, and sometimes a discharge from the nose. The coughing can precipitate prolapse during pregnancy. The small lungworm (hair lungworm) can cause pneumonia and bronchitis.

Good nutrition helps to build up resistance to the worm (as with other parasite species). Keep the sheep away from ponds and wet areas where snails can be found, as several species of slugs and snails act as intermediate hosts for the lungworms. When an infected sheep coughs, eggs are expelled and eaten from the grass by other sheep. This is something to consider when buying sheep from a farm having low-lying pastures.

Tramisol® given once a year should control lungworms. *Ivomec®* and Fen-benzadole are also effective.

Tapeworms. The feeding head of the tapeworm injures the intestine, and is thought to facilitate absorption of the toxin involved in enterotoxemia. (If you have vaccinated with Covexin-8, then enterotoxemia will not occur.) Tapeworms are not usually the primary worm infestation in a sheep, but since the passed tapeworm segments are large enough to be seen on the sheep droppings, their presence is alarming. Fenbenzadole (Panacur®, Safegard®) is effective for tapeworms. A moderate level of tapeworm is said to be of little damage to adult animals, but can seriously retard the growth of lambs.

Nose bots. The nose bot, *Oestrus ovis*, is a fly in its mature form, dark gray and about the size of a bee. The full-grown larvae are thick yellowish white grubs

Oestrus ovis larvae in horn cavity of sheep.

about one inch long, with dark transverse bands, and found primarily in the frontal sinuses of sheep. When deposited by the fly on the edge of the nostril, the grub is less than 1/12 of an inch long, and gradually moves up the nasal passages.

During fly season, sheep will put their heads to the ground, stamp, and suddenly run with their heads down, to avoid this fly. They often become frantic and press their noses to the ground, or against other sheep, as the flies attack them. This is usually during the heat of the day, letting up in early morning and late in the afternoon, and is more prevalent in areas with a hot summer.

The head grubs cause irritation as they crawl about in the nostrils and sinuses, and the resulting inflammation causes a thin secretion, becoming quite thick if any infections are present. As the mucous membranes are affected, and the secretions thicken, the sheep has difficulty in breathing and may sneeze frequently. They can become run-down because they lack appetite, or are so annoyed by flies that they cannot graze in peace.

The older treatments, injecting or spraying creosol solutions into the nasal passages, were barbaric and ineffective, and are no longer used. IvomecR has a slow effect on nose bots, taking up to thirty days before they are all dead, decomposed, and sneezed out.

Liver flukes. The three kinds of liver flukes all require an intermediate host; that is, part of the life cycle of the parasite is perpetuated in another creature. In the case of flukes, it is a snail or slug, found on wet marshy land.

Ponds, ditches, or swampy land provide the breeding place for the snails, so this kind of pasture is not ideal for sheep. If possible, drain wet areas where snails propagate, or fence off the marshy parts. Snail-destroying chemicals can be used, if the area does not drain into waters having fish or used for drinking water for humans or livestock. Most chemical treatments contain copper sulfate and would poison the pasture for sheep.

Liver fluke snails can be destroyed with less copper sulfate than has been used in past years. Mix 1 pound of copper sulfate snow with 4 pounds of sand, then apply 13 ounces of the sandy mix per acre. The snails emerge from their muddy homes on a regular schedule, so twice-a-month sanding should give good control. Using more than 13 ounces won't give any longer control because the copper is inactivated by the organic matter in the pastures.

As with other forms of parasites, liver flukes sometimes cause bottle jaw, or pot-belly, during the earlier stages, followed by loss of condition, diarrhea, weakness, and death. It can be diagnosed accurately in the liver of a slaughtered sheep, and sometimes can be diagnosed by microscopic examination of feces.

Merck's Curatrem® (clorsulon), while not yet cleared for sheep, can be prescribed by your veterinarian. It does kill developing flukes as well as adult, and is especially helpful on those animals that are treated in early stages of infestation.

Coccidiosis. Coccidia are microscopic protozoan parasites, present in most flocks without causing any problem. Outbreaks of coccidiosis are mainly in feedlot lambs of age one to three months, being raised in crowded conditions, but seldom in the pasture arrangement of a farm flock. Any rapid change of feed ration may predispose the lambs to an outbreak of coccidiosis, which usually appears within three weeks of the time they are brought into the feedlot. Other factors are chilling, shipping fatigue, and the interruption of feeding during the shipping time.

Small amounts of coccidial oocysts may be found in most mature sheep, but they seldom show any symptoms of infestation. However, these apparently healthy sheep may be carriers and contaminate their surroundings so that lambs, which are weakened by any change in ration, may by susceptible.

To prevent this, lambs should be fed during shipping, and should not have their ration changed too abruptly from grass to whatever concentrated feed will be given them. Overcrowding and contamination of feed and water must be prevented, for this is the main source of infection.

It once was believed that each species of animal had its own type of coccidia, and there was no cross-infestation. Later experiments have proven that some types of coccidia (there are a number of them even within those that infest sheep) are transmissible to different animal species, which then act as intermediate hosts. Some microscopic cysts in the muscle tissue of cattle or sheep, or even in intestinal tissue, can be fed raw to dogs, with the result that the dogs become infected and pass sporocysts. However, cooking or freezing will apparently render these parasites noninfectious, so meat fed to dogs or cats that associate with livestock should be previously cooked or frozen.

Symptoms. The symptoms of coccidiosis are diarrhea, then diarrhea with straining, chronic dark green or bloody diarrhea, loss of appetite, and some deaths. The lambs that recover are usually considered immune. A routine fecal sample showing evidence of this parasite will allow you to use the appropriate drug at an early stage.

Prevention and treatment. Once coccidiosis is diagnosed (or before, as a preventive), Bovated® (lasalocid) may be used, as it is an approved feed additive for sheep. Antibiotics can be given to prevent any secondary bacterial infection in intestines damaged by coccidia. Amprolium in the form of 1.25 percent crumbles can also be prescribed and fed for twenty-one days during an outbreak. Since sicker animals will not eat as well, they should then be separated and treated for another twenty-one days.

Lasalocid, the active ingredient in Bovatec, has produced no coccidial resistance problems, allowing it to be used for long and effective protection, and withdrawal of Bovatec-medicated feed prior to slaughter is not required.

WORM RESISTANCE TO DRUGS

To compound the problem of parasite control, there has been the development of resistance to some worm-control products in selected worm species.

Contrary to popular opinion, worms, insects, and bacteria do not alter their genetic makeup and become resistant to the particular drug or insecticide simply because they have been exposed to it—in the manner that one's skin develops calluses (resistance to blisters) by exposure to a shovel handle. All populations of living things contain individuals that naturally possess a greater resistance to something than the average population. For example, if we were using a fictitious wormer called "ATZ," it would kill all the ATZ-susceptible worms, but those few individuals that were naturally tolerant of ATZ would not be killed. Their offspring would carry the resistance factor, and if we continued to use ATZ to worm the sheep, the percentage of ATZ-resistant offspring in the population would increase each generation until the majority of the population was ATZ resistant. Resistance is a "selection" process rather than one of development. It would be like having a drug that prevented white sheep from conceiving, but black sheep were naturally "resistant" to it. If we fed that drug to a mixed black/white flock, sooner or later the majority of the flock would be black, not because the black sheep "developed" resistance by their exposure to the drug, but because they were "selected" from the population.

Possible wormer resistance increases with the frequency of treatment because we keep killing off the susceptible worms and leaving the possible resistant ones to regenerate the population. However, the development of resistance to a particular wormer is not a cut-and-dried foregone conclusion, any more than every species of bacteria develops resistance to a particular antibiotic. Some species do and some species don't. Do not assume that just because you use a wormer that you are necessarily developing resistance to it. On the other hand, you should not routinely worm on a calendar basis if it isn't necessary.

On farms where worms are only an occasional hazard—such as when pastures are obviously understocked, sheep and cattle are grazed alternately, and pastures are rotated—treatments are probably needed only when parasitism becomes evident. In other situations, regular worming on a three- to four-week schedule may be necessary.

Keep in mind that if you must worm very frequently, you are increasing selection pressure on the worm population, and resistance may become a problem sooner for you than for your neighbor who needs to worm less often. The answer to the question "How often do I need to worm?" is the same as to the question "How often do I need to wash my car?" When necessary!

The old recommendation to change wormer drugs often to avoid developing resistance is wrong. In fact, evidence suggests that this procedure may be the way to cause the worm population to become resistant to everything in the shortest period of time. It is now recommended to use the same wormer until you see resistance becoming a problem, then switch wormers. My veterinarian has used the same product in his flock for the past four years without any suggestion of resistance development. (He's in Kansas and some years must worm the entire flock every thirty days.) If you must change wormers, do not alternate with a wormer of the same chemical family or class. Resistance usually develops on chemical class lines, not brand names. Read the fine print for the generic name or chemical class. If you can't figure it out, ask your veterinarian or county farm advisor for help.

The more effective a wormer is on all the different species (broad-spectrum) the less chance of selection for resistant strains. Do not underdose the wormer, because natural resistance is rarely an all-or-none phenomenon; it is a dose-related thing. By underdosing, you may allow a *marginally* resistant worm to survive and propagate offspring with greater natural resistance, when it might have been susceptible to the full dose. With a highly effective drug, the worm numbers become so depleted that they lack the genetic variability required for selection for resistance in a short time.

To know for certain whether you have worm resistance to the drug you are using, you need egg counts. If egg counts are taken just before and then one week after administration of the correct dose, and the percentage decrease in egg count is less than 80 percent, the presence of anthelmintic-resistant parasites must be strongly suspected.

Researchers at the University of Massachusetts suggest that wormer-resistant nematodes are either unwittingly purchased, or they are produced (selected for) as a result of regular long-term use of the same wormer, or by underdosing. To avoid introducing resistant strains, you would need to treat all incoming new sheep. They could be given a double dose of two different wormers. As resistance to benzimidazoles is rather widespread, the two wormers to be used are levamisole and Ivermectin (prescribed by your veterinarian as it is not yet approved for sheep use). After treatment, the animals should be penned in a dry grass-free area for twenty-four hours to avoid contamination of pasture with viable nematode eggs that have not already passed out of the sheep when the worms are killed.

External Parasites

SHEEP TICKS

THE SHEEP TICK is not a true tick, but a wingless parasitic fly, known as a *sheep ked*, that passes its whole life cycle on the body of the sheep. It lays little brown pupae which are white inside. These hatch into almost-mature keds in about nineteen days.

Ticks are blood suckers, and roam all over the sheep, puncturing the skin to obtain their food. This causes firm dark nodules to develop and damages the sheepskin, reducing its value. These defects are called "cockle" by leather dealers. Cockle was at one time thought to be possibly a nutritional problem.

The ticks can produce such irritation and itching that sheep rub and scratch and injure their wool, and bite at themselves to relieve their suffering, sometimes becoming habitual wool chewers. They may get impacted rumens from eating the wool. Ticks reduce weight gain by causing varying degrees of anemia, and impair the quality and yield of the meat. The wool value is lower, as the ticks stain the wool with their feces, and the color does not readily scour out. Wool that is tick stained is sometimes referred to as "dingy."

Ticks can be easily eradicated with systematic treatment. The mature tick lays only a single puparia a week, for a total of a dozen or so in her lifetime. The pupa shells are attached to the wool about 1/2 to 1 inch from the skin. Therefore most of them are removed in shearing, making it easy to eliminate ticks by treating after shearing. The newly hatched ticks die within an hour unless they can suck blood from a sheep. The mature tick cannot survive more than two to four days away from the sheep.

To be effective, all sheep must be treated for ticks at one time, otherwise the untreated ones will pass the ticks back to the treated ones. Examine a new lamb or sheep before turning it in with your own, and treat it if you find even a single tick.

In the nineteenth century, the adult sheep were seldom treated for ticks. Since the shearing was done later in the spring than is common now, the heat of the sun and the scratching of the sheep drove most of the ticks onto the nicely wooled lambs. Herders waited a few weeks after shearing, then dipped the lambs in a liquid tobacco dip, sometimes with soap added. The vat used was a narrow box, with a slatted

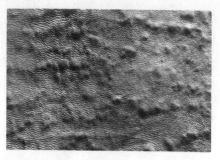

Sheep ticks and what they do: Photo at left shows closeup of keds and puparia including delivery of a single larva. At right, surface of pickled sheepskin showing cockle as firm, dark nodules. (USDA, Agricultural Research Center, Eastern Regional Center, Philadelphia)

grooved shelf at one side. The lamb was lifted out and laid on the shelf. Then the workmen squeezed the fleece, letting the dip run back into the box. By reusing the dip, five or six pounds of cheap plug tobacco could treat 100 lambs, and was quite effective on the ticks, although the mature sheep still had enough ticks left to get a good start on the next infestation.

Don't make the mistake of leaving any of your sheep with ticks. Every sheep must be treated in one session.

DIP

The standard method in very large herds is to run all the sheep through large dipping vats full of sheep dip liquid, or through spraying vats, where they are given a high-powered spray from several sides at once. This is done usually ten days after shearing, while the wool is still short, but after shearing nicks have healed.

For a small farm flock, this method is hardly practical. It is also unnecessary, for an efficient job of de-ticking requires practically no equipment at all.

SPRAY OR SPRINKLE

Low-pressure sprays, from 100 to 200 pounds per square inch, are ideal for treating sheep when they have been sheared recently and the wool is short.

Sprinkling, with insecticide solution in a garden sprinkler can, requires very little equipment.

SLOSH

Sloshing wets the animal thoroughly while avoiding many of the disadvantages of spraying and dipping.

The sheep do not contaminate the solution with germs that cause infection in a shearing nick, as they are not immersed in the liquid. It is dipped by bucket from a large container of solution and sloshed onto them. We lay them down, roll them on their backs for a good wetting-down of their bellies and undersides and necks.

Next, we stand them up and pour the liquid over their backs, giving special attention to the neck region where ticks and pupae are more concentrated. Then we stand back while they shake themselves.

Whatever chemical (report on chemicals to follow) or method you use, if the sheep are badly infested you should repeat the treatment in twenty-three or twenty-four days. If you treat them sooner, you may miss some of the unhatched pupae, and if you do it later, the unhatched ones have time to hatch and mature and lay their first eggs.

POWDER METHOD

Some de-ticking chemicals can be used to powder sheep, for tick control. It is a good method for the follow-up, twenty-four days after sloshing or spraying. You can thus virtually eliminate all ticks within a short time, with these two treatments, and not have to do it every year—unless you bring in a new sheep, or borrow a ram who has ticks, or loan out your ram to someone whose ewes have ticks. If you loan out your ram, treat him for ticks before returning him to your flock, and you will not have to treat everyone.

INJECTION

Ivomec (Ivermectin) wormer, which is injectable, is effective against most internal and external parasites, including ticks. It is not effective against tapeworms, flukes, or biting lice. As it is approved for other species, approval for sheep is anticipated. Your veterinarian can prescribe it.

Partially clipped skin showing keds and puparia at the base of the wool. (USDA)

SHEEP KED CONTROL CHEMICALS

Ectrin, Expar, Atroban. These are synthetic pyrethroids (a stable form of the garden insecticide made from the chrysanthemum). They are the product of choice because they are not a systemic that enters the body, and are considered both safe and effective. Used for both ked and lice control as a pour-on, spray, or ultralow-volume spray. No label withdrawal period prior to slaughter.

Ivomec. This wormer is also effective against keds (ticks) and against sucking lice. Consult label for withdrawal times.

Rotenone. See special paragraph to follow.

Co-Ral (Coumaphos). A systemic organophosphate. Used as an 0.6 percent spray or dip, or a 0.5 percent dust (1 to 2 ounces per sheep). Do not use on lambs under three months of age. Consult label for withdrawal time.

Diazinon. Used as an 0.5 percent spray or dip. No slaughter withdrawal. Not to be used on lambs less than one month old.

Government regulations on chemicals can change from year to year, as new chemicals are approved, tolerances change, and old chemicals are sometimes banned. Your agricultural extension agent can usually give you a printed list that is current.

ROTENONE

At one time rotenone was listed in the USDA information pamphlets as "safe and efficient for elimination of sheep ticks with one dipping," but then these external parasiticides became regulated by the Environmental Protection Agency. When EPA requested additional data and studies on the safety and effectiveness of all pesticide products, the manufacturers of rotenone did not desire to go to the expense to retest and tabulate and furnish this information, according to Dr. Eldred E. Kerr. Dr. Kerr, the western regional veterinarian for the FDA, suggested this was due to the competition from the many synthetic chemicals. Anyway, by default, rotenone is not now "approved" for official use. It is still an acceptable insecticide for garden vegetables, up to the day you eat them. Being the powdered root of a tropical plant, rather than a manufactured chemical, and an insecticide used by organic gardeners, the sheep raisers who have used it so successfully in the past will probably continue to use it.

For anyone interested in how it was used when still "approved" for use:

Rotenone

For dip or slosh, use 8 ounces of the 5 percent wettable powder to 100 gallons of water (about 1 1/2 ounces in 20-gallon garbage can). Mix it to a paste in a small amount of water and add it to the large quantity of water in the can, stirring well. The addition of a small amount of liquid dishwashing detergent will make it more penetrating and effective. Safe for ewes and lambs, up to the time of slaughter. For dusting, use 1.5 percent (garden type) dust, about 2 ounces per sheep.

The effect of rotenone is not immediate, so don't be alarmed if ticks are still moving a few hours afterward. They are no longer biting, and will die.

WOOL MAGGOTS
(Fleeceworms)

The maggot problems that are seen in most parts of the country are not the real screw-worms that were so destructive in Texas and some other places, but must be contended with in all warm summer weather.

Several types of blowflies lay maggot eggs, and all are about twice the size of house flies. They appear in the spring, and reproduce from then through hot weather, laying their eggs in masses at the edge of a wound, or in manure-soiled fleeces. The eggs hatch in six to twelve hours, and the larvae feed on the live flesh at the edge of the wound. They enlarge it and can, if not detected, eventually kill the animal.

Maggots need not necessarily be a big deal; the main thing is overcoming one's own revulsion. You can get rid of them quite easily, if they are not too advanced in damaging the sheep. The real danger is not knowing they are there. You can needlessly lose an animal if you do not look at your sheep, catching and examining them if anything looks at all suspicious.

Watch for moist fleece areas, or any injury that may become infested. Notice if animals scratch excessively on fences. This could be maggots or ticks.

When you locate an infestation, clip the wool around it and spray it with Screwworm Spray, or Wound Guard Smear, or any of the fly-strike aerosol bombs sold in livestock-supply catalogs (see Sources).

If you don't have one of these sprays, pick out all the maggots you can see and disinfect the wound. One of the sheep tick chemicals can substitute for fly repellent if necessary. Even if sprayed with repellent, the sheep should be kept under observation for a few days and treated again if needed. If the sheep have not been sheared, you might want to shear them after treating the area and removing all the maggots. That would make it easier to spot other infestations. Maggots often infest dog bites, and if your sheep are chased by dogs, check them often for unnoticed wounds and fly strikes.

The wool maggot or fleeceworm (maggot of the blowfly) can be distinguished from the even more dangerous screwworm (now eradicated from the U.S.). You can see the wool maggots move and crawl around, while the screwworms did not move, since they were imbedded in the flesh of the sheep.

PREVENT MAGGOTS

The following measures will lessen the chances of trouble with maggots:

1. Keep rear ends of ewes regularly tagged, especially any time that droppings become "loose" due to lush pasture or stomach worms. Worm your sheep regularly. Urine also attracts blowflies if it soils heavy tags.
2. Treat all cuts or injuries or shearing nicks with fly repellent during hot weather. Injuries or even insect bites invite flies.

3. Put fly repellent on docking and castration sites on lambs in warm weather. Check them periodically until healed. You can avoid the problem, in this instance, by having lambs early in the spring, before hot weather.
4. Use fly traps or large electronic bug-killers, to cut down the number of flies in your barn area.
5. Be especially vigilant during prolonged wet periods of summer. Warm and moist conditions favor the development of situations suitable for fly strike.

If you have a wool maggot problem with your whole flock, which is unlikely unless they have been attacked by dogs, you can use one of the sheep dip chemicals on all of them. I would probably use Atroban or Expar.

COMMON SCAB MITE

Several kinds of parasitic mites produce scab in sheep. The Psoroptes ovis is the common scab mite, a pearl gray mite about 1/40 of an inch long, with four pairs of brownish legs and sharp pointed brownish mouth parts.

The mites puncture the skin and live on the blood serum that oozes from the punctures. The skin becomes inflamed, then scabby with a gray scaly crust. The wool falls out, leaving large bare areas.

This is not to be confused with the loss of wool that sometimes occurs along the backbone of some breeds of sheep, when kept in areas of heavy rainfall.

To determine whether mites are present, scrape the outer edge of one of the scabs (the mites seek the healthy skin at the edge of the lesions) and put the scrappings on a piece of black paper. In a warm room under bright light, examine the paper with a magnifying glass. The mites become more active when warm, and are visible under the glass.

The common scab mite, often called "mange mite," is still a reportable disease in most states, but has all but been eradicated in sheep. However, should an infestation occur, it is very susceptible to Expar, Atroban, or Ectrin. These are effective as a spray or slosh, and can even be used on pregnant ewes. No withdrawal period before slaughter with these three medications. There are other pesticides that have been used in the past, requiring long withdrawal periods.

All sheep must be treated in one session, for the mite is quite contagious from sheep to sheep. Infected premises should not be used for clean sheep for thirty days.

LICE

Lice are probably second to ticks (keds) among the common ectoparasites affecting sheep. One species of biting lice and several species of sucking lice affect sheep. The eggs are attached to the individual wool fibers and hatch in one to two weeks into the nymph stage. After several molts which require another two to three weeks, the nymphs emerge as adults.

The feeding lice cause intense irritation and itching to the sheep, which results in restlessness, constant scratching and rubbing against walls and fences, interrupted

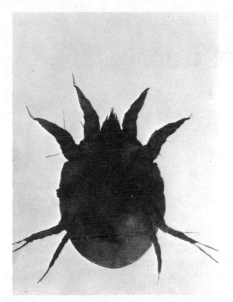 

Psproptes ovis, the common scab mite.
(USDA)

Sheep ked. (USDA)

feeding, loss of weight, and severe damage to the wool. The cardinal sign of lice in the flock are hundreds of telltale tags of wool hanging from fences, trees, etc., where the sheep have been rubbing.

Lice are very susceptible to the commonly used insecticides, but often two treatments are needed to kill any newly emerged nymphs (as the egg is a protected stage). Once removed from the sheep, they will not return until you introduce more lice-infested sheep into the flock. If in doubt, you should treat any new animals prior to placing with your flock, to prevent the reestablishment of a lice or tick infestation.

The pyrethroid products (Atroban, Expar, Ectrin) do an excellent job of ridding the flock of lice, and are very safe products to use.

Hoof Care

IF YOU BUY SHEEP that have hooves in good shape and recently trimmed, they should need trimming only about twice a year.

Many foot disease conditions can be prevented by proper and periodic hoof trimming, most easily done in the spring when hooves are still soft from wet weather, and in the fall after the start of a rainy season. The amount of hoof wear depends on whether soil conditions are mud, sand, or gravel, and whether the barn has a dirt or concrete floor. In some situations, hooves may need trimming more than twice a year, especially when the weather is wet for prolonged periods.

LAMENESS

POSSIBLE CAUSES OF LAMENESS

- Overgrown untrimmed hooves.
- Wedges of mud, or stone, or other matter lodged in the cleft of the hoof.
- Plugged toe gland. Squeeze to remove plug, then disinfect injury.
- Sprain, strain, nail puncture, or thorn.
- Abnormal foot development. Genetic defect, cull out.
- Foot abscess.
- Foot scald.
- True infectious hoof rot.
- Vitamin deficiency. Try ADE vitamin in food or injection.

You can help prevent sheep from becoming lame by:

1. Trimming all feet each spring prior to new pasture.
2. Trimming again at shearing time or later in the year. Untrimmed hooves curl under on the sides, and provide pockets for accumulation of moist mud and manure, ideal for growth of foot disease germs.
3. Maintaining dry bedding area during winter.

Here are the three steps to put a sheep down for hoof trimming: (Upper left) Slip your right thumb into sheep's mouth, back of the incisor teeth, and place the other hand on the sheep's right hip. Bend sheep's head sharply over its left shoulder as you press your hand down and swing sheep toward you. (Upper right) Lower sheep to the ground as you step back. From this position you can lower her to the ground completely, for shearing. (Right) Sheep is raised on rump for hoof trimming.

4. Keeping sheep away from marshy pastures during wet months.
5. Changing location of feeding sites occasionally to prevent accumulation of manure and formation of muddy areas.
6. Having footbath arrangement for use when needed.

CHECK LIMPING SHEEP

When you notice a sheep limping, try to discover the reason. Notice which foot is being favored, then catch the sheep and trim all four hooves if they need it, doing the sore one last, so as not to spread any possible infection.

Photo at left shows hoof edges curling under. Don't let them get worse than this before trimming. At right, they are properly trimmed. The trimming job is easy if you do it twice a year.

HOOF TRIMMING

Using a hoof knife or jackknife, trim the hoof back to the level of the foot pad, so that the sheep can stand firmly and squarely on both claws. The purpose of trimming, other than to prevent lameness, is to give a good flat surface on the bottom, with both pads of the hoof evenly flat. To do this, trim off the excess horn so that it is level with the sole, and also not protruding too far in front. If there are still any pockets where mud and manure can gather, dip these out with the point of your knife or the hook on the end of the hoof knife, and trim the hoof back a little farther. Notice the shape of the hooves on your half-grown lambs, for the ideal.

Hoof knives are sold in two sizes, large for cows and smaller for sheep. In dry weather, when feet are drier and harder to trim, hoof shears can be useful for part of the job.

In hoof shears, one of the best to appear recently is a Swiss-manufactured pruning shears (Felco –2), which has come into routine use in many large commercial flocks. The curved blades have less tendency to slip on tough dry hoofs. They are slightly more expensive than the traditional Burdizzo shears, but more than worth the extra money. The specially tempered blades are thin and very sharp, requiring less than 25 percent of the "squeeze" power needed with the more traditional shears. Because of their sharpness and ease of use, exercise caution when first using these, because it is very easy to overtrim the hoof and/or cut your hand, even if you are experienced. ALWAYS wear a leather glove and arm protection on the opposite hand when trimming hooves to avoid accidental injury should the sheep kick.

When trimming hooves of a sheep that has been limping, look for any lump of mud, or a stone or sharp splinter between the claws of the hoof that seems to be sore.

FOOT GLAND

If there is nothing there, check the gland. Sheep have a deep gland between the two toes of each foot, with a small opening at the front of the hoof, on top. This can be readily seen if you look for it. Goats do not have these. The gland's secretion is waxy and has a faint, strange odor, said to scent the grass and reinforce the herding instinct.

If these glands become plugged with mud, the secretion is retained and the foot becomes lame. Squeeze the gland, and sometimes a fairly large blob of waxy substance pops out. If this was the problem, then the sheep should improve.

While on the subject of glands, there are two on the face, just below the inner corner of the eyes. The fatty secretion from these is sometimes noticeable and looks as if it is coming from the eyes. On dark (black sheep) of the Lincoln breed, there is usually a pale gray or white marking at these glands. There are also two mammary pouch glands in the groin, whose scent attracts newborn lambs to the udder.

If there is no evidence of a plugged foot gland, or a foreign object between the toes, try to determine if a hoof disease is present. You will have to get a clear idea of what a normal hoof looks like, before you can spot a diseased condition. If you're not familiar with sheep hooves, compare the sore one with another of her feet.

If you are unable to find the cause of the lameness, you can play safe by trimming hooves on all of the sheep. They may need it now anyway. Squirt the hooves with a disinfectant, or one of the footbath formulas to follow.

We find a foot rot spray medication for cattle, recommended to "fight bacteria" in general, to be very useful for foot irritations. It contains hexachlorophene, dicholorophene, silicone (to hold on the medication) and alcohol. It is available from sheep suppliers. (See Sources.)

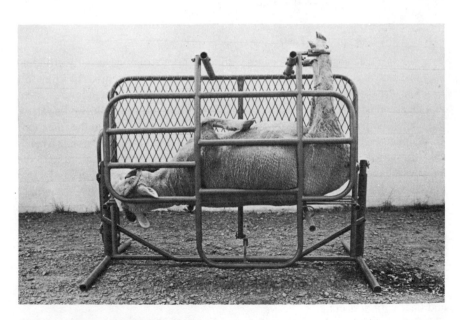

MIGHTY MIKE SHEEP SQUEEZE
Sheep enters squeeze and sides are squeezed together. It is then simple to turn sheep completely upside down. Feet can be restrained in stirrups for foot trimming. Adjustable from 80-pound lamb to largest adult ewe or ram. (See Sources.)

FOOT SCALD

Foot scald is sometimes mistaken for foot rot. In scald, the soft tissues above and between the toes are involved. There is inflamed tissue and moistness, and sometimes open sores, usually involving only one foot. It is very similar to athlete's foot in humans.

Foot scald is caused by dampness, wet pastures, prolonged walking in mud, or the abrasion of dirt or foreign objects lodged between the toes. The soft tissue between and above the toes and "heel" become irritated and inflamed. This occurs primarily during the wet winter and spring months, and the condition sometimes improves without treatment, in dry weather. It is a major problem only as it lessens foot resistance to more serious disease like abscess or foot rot, and causes lame sheep to eat poorly and not get enough exercise.

Treatment. Trim hooves and spray with antibacterial hoof spray. If no improvement, treat with footbath, the same kind of zinc sulfate as for hoof rot, or ordinary hydrogen peroxide.

If you do not have footbath facilities, use a large fruit-juice can, filled with 2 inches of the footbath solution, and soak the affected foot for five minutes. Repeat if necessary.

Penicillin injections may be helpful.

Prevention. Since foot scald is primarily caused by dampness and mud, it helps to get rid of muddy places if possible.

FOOT ABSCESS (Bumble Foot)

This is a true abscess, and occurs within the hoof structure, usually afflicting only one foot. It is considered infectious, but not extremely contagious like foot rot.

The infection causes formation of thick pus, and as the internal pressure increases, the sheep becomes more and more lame. Sometimes you can see a swelling above the hoof. Compare it with the other foot—it will be warmer due to the infection.

It is caused by bacteria in manure and dirt, which enter through cuts or a wound, causing an infection of the soft tissue. There is usually a reddening of the tissue between the toes. This infection may become advanced if not treated, and move into the joints and ligaments. At that stage it is almost incurable because it can't be reached.

Abscess is dangerous in pregnant ewes as they will fail to graze, be slow about getting to grain feeding, and not get enough exercise, which can bring on pregnancy toxemia. Insufficient nutrition also leads to low birth weight of lambs, and having insufficient milk for them.

Treatment. Unless pressure is released by an incision, the abscess may eventually break and discharge pus. When it is opened or breaks, squeeze out the pus and treat with antiseptic.

New Zealanders clean out the infected area, treat with footbath, and bandage the foot. They follow this with an intramuscular injection of up to a million units of penicillin. Penicillin treatment is usually continued for from three to five days.

FOOT ROT

Sheep raisers once thought that foot rot was a spontaneous disease of wet weather. It was only about fifty years ago that the primary causative bacteria were identified, and found to be capable of surviving less than two weeks on contaminated pasture.

Foot rot is caused by a bacterium called Bacillus nodosus (present in more than 20 different serotypes in the U.S.). Clean sheep become infected by walking over ground contaminated by infected sheep within the previous seven to ten days. While the bacterium does not survive on the ground much longer than seven days, it can survive indefinitely in the feet of infected sheep. Its spread is especially rapid in warm moist weather.

The foot rot organism is an anaerobe, which means it grows in an oxygen-free environment—deep in the hoof tissue where little or no oxygen is present. This is why hoof trimming is an important part of foot rot treatment, so that dead tissue is removed to allow oxygen to enter.

The availability of Footvax vaccine, coupled with hoof paring and hoofbath solutions, makes both prevention and cure possible.

Symptoms. Foot rot starts with a reddening of the skin between the claws of the hoof. Odor is faint or absent in the beginning, as it is caused by destroyed tissue. The infection starts in the soft horny tissue between the hoof, or on the ball of the heel, then spreads to the inner hoof wall. By this time it has developed a strong unpleasant odor. As the disease progresses, the surface of the tissue between the under-run horn has a slimy appearance. The horny tissue of the claws becomes partly detached, and the separation of the hoof wall from the underlying tissue lets the claw become misshapen and deformed. There is relatively little soft tissue swelling. In severe infections, it is often more practical to dispose of the most seriously affected animals and concentrate treatment on the milder cases.

Treatment.

1. Hoof trimming, removing as much of the affected part as possible, and exposing infected areas to the footbath. Disinfect knife after each hoof trimmed. Burn the hoof trimmings.
2. Footbath, formula to follow.
3. Hold sheep on a dry yard or pasture for twenty-four hours, if possible, after footbath treatment.
4. Vaccinate with Footvax. See following section on vaccine.

TREATMENTS

FOOTBATH

If you run the sheep through a trough of plain water first, it keeps the bacterial bath clean longer. Be sure that sheep have had water and are not thirsty, so they do not drink any of the footbath.

Feet should be trimmed before the footbath, not just to allow better penetration, but because foot preparation will harden the hooves, making them more difficult to trim. Remember to disinfect knife between each hoof and each sheep, so you do not needlessly spread germs.

The traditional footbath, in which the sheep were walked through the solution, was designed primarily for formalin footbath solutions in which the exposure time to the solution was purposely limited in order to avoid burning the soft tissue of the feet. Recent research with zinc sulfate footbath solutions has shown much greater (ten-fold or more) benefit if the animals are allowed to stand in the solution for one hour on two occasions spaced about a week apart. Do NOT attempt this with formalin or copper sulfate, as you could severely burn or "pickle" sheep's feet.

Trim nonlimpers first, then put them in the footbath first, and then turn them into clean pasture (not grazed for at least a week) if you have one. Next foot-bathe the limpers, and keep them in a dry area if possible, treating them regularly every five to seven days, or have them walk through the bath on the way to daily feeding.

An alternative to the standard footbath trough can be constructed out of a 4 x 8 sheet of 1/2-inch waterproof sheathing plywood, with 2 x 4's nailed around the edge. A little caulking will make it watertight. A temporary pen of lambing panels around the perimeter completes the unit, which will hold upwards of eight to ten animals or more, depending on size. In conjunction with vaccination with Footvax, trimming, and isolation of the infected animals from the clean group, total eradication of foot rot from treated flocks has been accomplished.

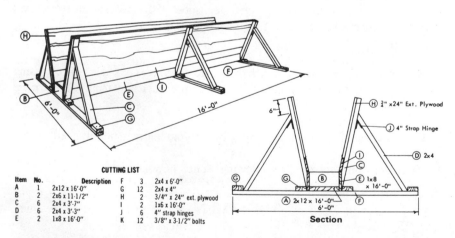

CUTTING LIST

Item	No.	Description			
A	1	2x12 x 16'-0"	F	3	2x4 x 6'-0"
B	2	2x6 x 11-1/2"	G	12	2x4 x 4"
C	6	2x4 x 3'-7"	H	2	3/4" x 24" ext. plywood
D	6	2x4 x 3'-3"	I	2	1x6 x 16'-0"
E	2	1x8 x 16'-0"	J	6	4" strap hinges
			K	12	3/8" x 3-1/2" bolts

Footbath trough is useful if hoof problems develop, and the footbath is combined with proper trimming. This trough need not have sides made of plywood. Ship lap or 1 x 6s can be used. A gate is needed at the exit to keep sheep standing in the bath for the required time. (Midwest Plan Service)

FOOTBATH FORMULAS

Copper sulfate (bluestone) has been widely used in the past, but is highly toxic and not used now. Formalin is also discontinued in general usage, as it is very irritating when inhaled, irritating to the skin and feet, and tends to excessively harden the hooves to make future trimming very difficult. It is also quite painful to raw exposed tissue, and penetrates the hoof poorly. Another major drawback is that the solution loses strength rapidly when contaminated with organic matter.

ZINC SULFATE SOLUTION

This solution is the least toxic and the most effective of all footbath preparations. It need not be changed as frequently because it does not lose its strength from organic contamination. The addition of some liquid laundry detergent will make it more penetrating and aid in dissolving the powder in water. Adding some wool tags to the footbath will reduce splashing and discourage sheep from drinking the solution.

Proportion. Use 8 pounds of zinc sulfate to 10 gallons of water. Zinc sulfate is somewhat "lazy" about going into solution, so mix slowly, as rapid mixing could cause it to set up like concrete. Add a cup of liquid detergent into 5 gallons of water, then SLOWLY add the 8 pounds of zinc sulfate powder to the water, stirring constantly. Do not breathe the dust as it can be as irritating as inhaling horseradish. Add the other 5 gallons of water, stirring. The solution should be about 2 inches deep in the trough. Do not rush the sheep through a walk-through bath, because they might splash the solution on their udders.

When tested against the commercial Footrite solution, the homemade 10-percent zinc sulfate solution was equally effective at a fraction of the cost.

Paraformaldehyde. *Shepherd* magazine mentions a formaldehyde derivative, paraformaldehyde, carried by chemical suppliers as a bathroom disinfectant, which can be purchased in flake form. It is sprinkled on the ground around waterers on a regular basis about once a week, and has been used in other countries to halt the spread of hoof rot. It is still necessary to pare the hooves, and still desirable to footbathe them in zinc sulfate.

ALTERNATIVE TO FOOTBATH TROUGH

If you have only three or four sheep, it may not seem practical to build the footbath arrangement. You can use a large empty fruit-juice or coffee can, with footbath mixture to a depth of two inches. For each sheep, use it on the lame foot, and as a precaution also on trimmed healthy feet, if you feel the infection is spreading. Apply the footbath with a brush to the hoof, then hold the infected foot in the can of solution. If you are treating a front foot, hold up the other front foot, forcing the animal to stand on the foot that is in the bath. Keep the animal on a dry floor a minimum of a half-hour before turning out to pasture, preferably a dry pasture. Repeat in a week, if still limping.

TOPICAL MEDICATION

These are preparations that are painted or sprayed onto the hooves just after trimming, when footbath facilities are not available. The prior trimming is most important, for the preparation to be effective. Some used are:

- 10 percent zinc sulfate in water.
- 10 percent zinc sulfate in vinegar.
- 2 parts copper sulfate in 1 part pine tar.
- 10 percent formalin in water.
- Penicillin in alcohol—5 million units of potassium penicillin G with 10 cc water, add to one pint of alcohol.
- Kopertox—this has long been one of the most effective of all topical medications for foot and minor wound problems. It is formulated with a vehicle that makes it stick to the tissue (as well as your fingers and everything else).

DRY CHEMICAL TREATMENT

Zinc sulfate, 10 percent in lime, can be placed in a box between feed and water, or spread on the ground around feeding troughs, to reduce the spread of hoof rot. This is more of a preventive than a treatment of existing disease, and is a convenient substitute for a footbath during freezing weather.

VACCINE

Sheep raisers are fortunate that the foot rot vaccine has finally become available. It is a 10-strain whole cell vaccine that contains all known serogroups of the foot rot bacteria.

To start out, you vaccinate all sheep on the property with Footvax. Six weeks later, all should receive a booster dose, followed by booster doses at four- to twelve-month intervals. All sheep's feet must be inspected, and animals affected with foot rot should be segregated. Misshapen feet should be trimmed to maintain a good shape, and those of infected sheep should be closely trimmed. The entire group should have a footbath of zinc sulfate. This can be regularly a few days apart for infected sheep, and at weekly intervals for those who appear to be clean. Severely infected sheep may benefit from antibiotics.

Any new sheep brought into the flock should be vaccinated with Footvax upon arrival.

This preparation is easy to administer. It is given subcutaneously just below and behind the ear, along the side of the neck. The timing of the vaccination should, ideally, be just prior to the season of greatest danger. This vaccine is 85 to 90 percent effective in preventing foot rot, and equally effective in curing active cases (when combined with foot trimming and footbaths). Its high level of effectiveness is due in part to the oil adjuvant (immunity enhancer). As with all oil adjuvants, a telltale bump will appear at the injection site, disappearing in a few weeks. Do not be alarmed if a few of these show a small amount of exudate after several weeks. Because of this bump, it is not advisable to vaccinate sheep just prior to showing.

CHAPTER 16

Medication

SUCCESSFUL TREATMENT of any sheep illness requires detection as early as possible, before the sheep is "down." With the development of new medications and antibiotics, it is no longer true that "a down sheep is a dead sheep," but the chance for recovery is much better if illness is diagnosed and treated before it has progressed. Some of the causes of illness are: lack of exercise, unsanitary housing, moldy or spoiled feed, toxic plants or other poisonous substances, improper diet (insufficient water and feed, or overeating), parasites, injuries, infection from assisted lambing, germs from other sick sheep, abrupt change of feed, and stress (weather, shipping, predators, etc.). Prevention is always better than treatment, and early treatment has better success than late.

HOW TO DETECT A SICK SHEEP

You must become familiar with the normal behavior of your sheep, even for each individual animal, to know when one is acting abnormally. Have some quick and easy way of catching them when needed, like a corral where they can be fed and then enclosed.

Signs of abnormality are loss of appetite, not coming to eat as usual, and standing apart from the group when at rest. Be concerned if a sheep is lying down most of the time, when the others are not. Any weakness or staggering, unusually labored or fast breathing, change in bowel movements or "personality," wool slipping, hanging the head over the water source, or temperature over 104 degrees is indication of possible illness.

The normal temperature of a sheep (except in very hot weather) is in the range of from 100.9 to 103 degrees (Average 102.3 degrees). A veterinary rectal thermometer has a ring or a hole at the outer end, where you can tie a string for easy removal.

If it is necessary to catch a urine sample, such as for use with the pregnancy toxemia (ketosis) strips of glucose strips for enterotoxemia, try a plastic cup fastened to the end of a shepherd's crook handle. Impatient? Try holding the sheep's nostrils closed for a moment. This "stress" sometimes triggers urination.

AREAS OF GERM TRANSMISSION

- Water or feed contaminated by feces from sheep or other animals can transmit intestinal diseases and certain parasites. Respiratory disease may also be spread by nasal discharge into drinking water or feeding containers.
- Dirty uncrotched wool on a ewe can infect the lamb.
- Manure accumulated in a lambing shed or around the feeding trough can intensify exposure to disease germs and coccidiosis, serve as breeding media for flies and other vermin, not to mention the production of ammonia fumes which can contribute to respiratory diseases.
- Feeding on bare ground, except through fence lines where food cannot be trampled or soiled by the sheep, greatly contributes to disease and parasite exposure.
- Wet muddy places predispose the sheep to hoof diseases.
- Newly acquired sheep can be carriers of many serious diseases such as foot rot and brucellosis.
- Venereal transmission of disease at breeding time.
- Insects, birds, snails, dogs, cats, and other hosts can be carriers of parasites and diseases.
- Dirty hypodermic syringes and needles can cause injection site infections and abscesses, and transmit certain infectious diseases.

Maintain sanitary surroundings. If you keep sheep adequately fed on a well-balanced diet, they are much less apt to eat poisonous substances, as well as being better able to withstand disease. If you worm them regularly (see Chapter 13, Internal Parasites) there is less of a parasite build-up to weaken them and leave them more susceptible to disease.

Sound management dictates that sheep should be vaccinated against the common ailments and diseases prevalent in your area.

DRUGS

Good shepherds are prepared for emergencies by having a supply of standard medicines on hand. These include bloat medication, antibiotics, propylene glycol for pregnancy toxemia, Cal-phos or other treatment for milk fever, iodine and disinfectants, mineral oil for constipation, dextrose solution, hoof footbath preparation, uterine boluses, and the proper clean or sterile equipment (syringes, needles, etc.) needed to administer them. Of the antibiotics, pen-strep* will halt many infections. For certain specific infections, other antibiotics are necessary.

Penicillin and tetracycline are quite safe to use, for their toxicity in sheep is extremely low. The severe allergic reactions that sometimes occur in humans are much rarer in animals. These drugs are of use in pneumonia, infection after lambing, and as a preventive against infection following cleaning and dressing of maggot infestation. They are of minor help for enterotoxemia (which could have been prevented by vaccination with Covexin-8).

*Penicillin-dihydrostreptomycin.

Many useful cattle products are not labeled for use in sheep. When you must use a medication meant for cattle, it is generally estimated that the drug can be administered at the same label dose per pound body weight basis. This usually works out that the dose for one cow would suffice for about five sheep. Since with many drugs the exact dosage is very important, with even a slight overdose being fatal, it is best to use sheep medicines when available, or seek the advice of your veterinarian, who can prescribe and obtain medicines not "registered" for sheep use (and also take the responsibility).

METHODS OF ADMINISTERING DRUGS AND VACCINES

1. Oral, by mouth such as worm boluses with bolus gun, or with capsule forceps.
2. Oral, powder such as vitamins, placed well back on the tongue for treatment of an individual animal, or in feed or drinking water for general treatment of whole flock.
3. Oral, liquid given as drench with syringe, or in drinking water.
4. Spray-on, such as pinkeye spray, or insecticides including maggot or screw-worm bomb.
5. Pour-on, such as iodine on newborn lamb navel, disinfectant on minor wounds, and certain insecticides.
6. Subcutaneous, medication injected just under the skin.
7. Intradermal, medication injected into the skin.
8. Intramuscular, liquid such as antibiotics injected into heavy muscle.
9. Pessaries, as uterine boluses to prevent infection after an assisted lambing.
10. Intramammary, injection of fluid or ointment through the teat opening, as mastitis drugs.
11. Intraperitoneal, injection of liquid through right flank into the abdominal cavity.
12. Intraruminal, injection of fluid into the rumen, on the left side, as for bloat remedy when too late to be given by mouth.
13. Intranasal, spraying of vaccine up the nasal cavity.
14. Intravenous, injection of fluid into a vein.

The last method of medication should obviously be done by a veterinarian or very experienced person. Numbers 11 and 12 are surgical procedures and should be done by an experienced person or certainly with the help of someone who has done them before.

ORAL MEDICATIONS

Boluses (large pills) intended for sheep will slip down easier if coated with mineral oil or cooking oil. Do not soak them or they will disintegrate. Just coat lightly before giving them. (See Chapter 13.) Boluses made for cattle should be given only at the recommendation of a veterinarian. Sheep may be able to swallow some cow boluses, but the wide diameter could cause them to lodge in the lower esophagus.

The easiest way to hold the sheep is to back it into a corner and straddle it, facing forward. You can hold the bolus in a "bolus gun" and eject it when you have the pill "over the hump" of the tongue and resting at the base. Some find it easier to deposit the bolus at the base of the tongue with a "capsule forceps" which are like a pair of pliers with long angled handles. In either case, exercise caution. The object is to *deposit* the tablet or bolus at the base of the tongue where the animal can swallow it, not jam it down the throat as if one were loading a cannon. Forceful jamming of the bolus too deep into the throat can cause the bolus to be deposited into the wind pipe (trachea) with fatal results. Also, it is easier to damage the throat with forceps than a proper balling gun. You will need to wedge the mouth open with the left thumb (in the space between the front teeth and the molars unless you want to be bitten) while gently inserting the bolus gun with the other hand. Don't release the sheep until you are sure the medication has been swallowed.

Liquid medication is given with a dose syringe, or a dose gun for larger numbers of sheep. The pipe or nozzle on your dose syringe should be about 5 to 6 inches long, and have a smoothly rounded tip that will not injure the sheep. The sheep's head should be held in a level position, with the nose no higher than the eyes, so that the liquid will not be forced into the lungs and cause pneumonia. The safest procedure is the "trickle" method. Administer the liquid slowly while holding the sheep's head in the level position.

Can cow medications be legally used for sheep? You often see the words, "not approved for sheep," "not cleared for sheep," "not licensed for sheep use." This can be explained because sheep are such a minor industry in this country that the expense of the required testing to pass FDA/USDA regulations is much too great to be warranted by the potential sale of the drug. Many of these drugs are approved in New Zealand and Australia, where sheep are big business.

Veterinarians are allowed to administer or prescribe drugs "off label" if they have been cleared for use in other species and/or no other effective treatment is available. It is essential to have the advice of a veterinarian on the dosage as well as on withdrawal times. There are quite a number of cattle and horse medications that are used safely and effectively on a regular basis for sheep.

INJECTIONS

General information. Sterile procedures must be maintained to avoid serious infections. Use only clean and sterile syringes (boiled at least twenty minutes if not using new sterile disposable syringes) and sharp sterile *disposable* needles. While needles can be boiled, this causes them to become quite dull. Dull needles are one of the greatest causes of injection site infections because they force dirt, grease, and bacteria to be carried through the skin. Storing needles in alcohol can also cause the needle points to be blunted by striking against the side of the container.

Disposable plastic syringes are inexpensive and can be ordered from veterinary-supply catalogs (see Sources) or in some states can be obtained from your local drugstore.

To fill a syringe with medication, first clean the top of the vial with a disinfectant to remove any dust or dirt. Swirl or shake the bottle to thoroughly mix the contents without causing undue bubbles. While holding the vial upside down, pull the syringe plunger back to approximately the volume of drug to be removed, insert the needle into the center of the vial stopper, and depress the plunger forcing the air into the vial. (This prevents creating a vacuum in the vial and difficult removal of the dose.) Withdraw a greater volume of drug than needed and then express the excess drug back into the vial to remove any air bubbles that may form in the syringe.

If you are withdrawing doses for a number of sheep, and particularly if you wish to save the balance of the contents of the medication vial, you can protect it from contamination by sanitizing the top of the vial with disinfectant as above, then inserting a sterile needle *which you will leave in the bottle*. Fill the syringe, leave the needle in the bottle, and attach a separate needle to the syringe for vaccinating. For the next dose, detach the used needle, fill the syringe with the needle left in the vial, leaving the needle in the vial, and reattach a new or disinfected needle for the injection(s). In this way, you protect your medication from any contamination, and can save the balance of the contents through the dating period. While this is true of an inactivated vaccine like Covexin-8, you cannot save a live vaccine (such as Nasalgen). Once opened and exposed to air, the live vaccines are no longer stable and cannot be stored for later use.

Once the needle is filled with medication, do not let it touch anything, or it will no longer be sterile. If possible, have a helper hold the sheep or hand you the necessary medicine and equipment.

An alcohol swabbing of the skin prior to injection will give the impression that the skin has been sterilized, but this is not really the case. It takes approximately six to eight minutes for alcohol to kill the common disease germs. The alcohol swab mechanically removes the majority of the skin bacteria contained in the body oils. Simple wetting of the skin with alcohol and most other disinfectants merely puts the bacteria in solution where they can be more readily picked up on the needle and carried into the injection site. My veterinarian assures me that injections placed in dry "clean" skin (free from excessive grease, manure, etc.) result in far less injection site contaminations.

For the same reason you should avoid, if at all possible, injecting wet sheep. Routine vaccinations should always be scheduled when the weather is dry.

Protect drugs from freezing and from heat. Many medicines require temperatures above freezing, and below 50 degrees. Read the label on each medication for storage directions. Many antibiotics require refrigeration. Check the expiration date on the package.

Read the dosage carefully and follow it, or use according to the advice of your veterinarian. On some drugs, there is not much leeway between the effective dose and the overdose that could be fatal or harmful.

Subcutaneous injection. Subcutaneous (or "sub-cu") means depositing medication directly between the skin and the underlying muscle tissue. The medication usually should be at body temperature, especially with young lambs, and can be given in the neck, but the preferred place is in the loose hairless skin behind and below the armpits, to the rear of the elbow, over the chest wall. Be careful not to inject into the armpit, as can happen if the injection is made too far forward. The armpit (axillary space) is actually a large cavity underlying the entire shoulder blade area, crossed by the major artery, vein, and nerves that serve the front leg. Some vaccines are highly

The correct method of sub-cu injection is to pull away a fold of loose skin, insert needle into the space under the skin, and avoid injecting into the muscle, near a joint, or in areas carrying more than a minimal amount of fat under the skin. The fold of skin is called the sub-cu "tent."

irritating, and if injected into the axillary space could cause severe irritation and lameness.

A dosage of more than 10 ml is best distributed among several sites instead of all in one place (use even less per site with lambs).

To inject, pinch up a fold of loose skin. Insert the needle into the space under the skin, holding the needle parallel to the body surface. Rub the area afterward, to distribute the medication and hasten its absorption.

Do not make the injection near a joint, or in areas carrying more than a minimal amount of fat under the skin. With this injection there should be little problem with veins, but if you want to make sure you are not in a vein, the plunger can be pulled out very slightly before injecting. If it draws out blood, try another spot. Medication for sub-cu use should never be injected into a muscle.

Intradermal injection (sometimes called intracutaneous). An intradermal injection is made into the skin instead of beneath it like sub-cu, and is rarely used. The inserted fine needle is so close to the surface that it can be seen through the outer layer of skin, in a site the same as for subcutaneous. Injection is made slowly while drawing out the needle, distributing the dose along the needle's course.

Intramuscular injection. An intramuscular injection deposits the medication deep into a large muscle, such as in the neck or heavy muscle of the thigh. If you can get an experienced person to demonstrate this, you can see the exact site that will avoid both a nerve and the best cuts of meat.

Fresh and sterile antibiotics and drugs are important, as are a sterile needle and sterile procedure to avoid risk of deep-seated infection. Use a new sharp disposable needle and syringe to avoid tissue damage. Sanitize the top of the vial stopper with alcohol before withdrawing the medication into the syringe.

With an assistant holding the sheep still, thrust the needle quickly into the muscle. To be sure it is not in a blood vessel, pull the plunger out a bit. If blood is sucked into the syringe, try another site. It is usually best not to inject more than a 10-ml volume into any one spot.

Intramammary. Infusion of liquids or ointments are sometimes administered into the teat for udder ailments such as mastitis. The nozzle or tube of udder antibiotics are designed for cattle, and are difficult to use in sheep.

Cleanliness is paramount when infusing the udder. First, milk out the affected side of the udder as completely as possible. Afterwards, wash your hands and the udder thoroughly, then carefully disinfect the teats several times a few minutes apart. A solution of Clorox laundry bleach is an excellent disinfectant. Dry the end of the teat(s) with a clean paper towel so that germs will not be carried in when you insert the medication.

Remove the cap of the infusion tube and gently insert it into the teat canal. Don't remove the cap until you are completely ready to use it, for you want to avoid bacterial or fungal contamination, which could complicate an already serious condition. Squeeze the dose into the teat, then massage the dose upward toward the base of the udder. Again, cleanliness cannot be over emphasized! Most udder infections can be helped by antibiotics, but unsanitary infusion techniques could introduce fungi and molds that are not sensitive to the antibiotic, resulting in a totally untreatable condition.

Intraperitoneal injection. This should be done only by the person familiar with aseptic technique and anatomy. Complications (peritonitis) are common after this procedure. If it must be done in an extreme emergency, have an experienced person guide you. It is easier if one person holds the sheep, straddling it just in front of the shoulders.

Clip the wool from the right flank, in the shallow triangular depression below the spine, between the last rib and the point of the hip bone. Medication injected into the center of this depression goes into the peritoneal (abdominal) cavity. Scrub the injection area with soap, rinse, dry, and then disinfect the skin with iodine.

Medication should be at sheep body temperature. A sterile 25-ml or 50-ml syringe and a sterile 16-gauge needle are required. Disinfect the bottle stopper before withdrawing the medication, and use a separate sterile needle to give the medication in order to reduce the possibility of introducing the infection into the body cavity. Hold the needle perpendicular to the skin, pointed toward the center of the body. Inject quickly the full length of the needle, and eject the medication. If it does not inject easily, the needle may be clogged with a plug of tissue, or may not be in the right place. If so, withdraw the needle, replace it with a new one, and try again. Rub the injection site with disinfectant afterwards.

ANTIBIOTICS

Antibiotics is the general term for a group of products that either kill or seriously impair bacterial growth. They are effective against many bacterial diseases, but are totally useless against diseases caused by viruses.

Antibiotics are only effective when present in adequate concentration. Low concentration (below recommended levels) or discontinuation of treatment too soon may fail to kill the more resistant bacteria present in the infection. This could result in a relapse of the condition or, more seriously, a chronic infection, which could be very difficult or impossible to treat due to bacterial resistance to the antibiotic.

The availability of antibiotics should not encourage improper sanitary practices or "fire-engine" treatment of diseases that can be prevented through proper management and vaccination. There is concern that improper use of antibiotics can give rise to new strains of drug-resistant bacteria that pose a threat to both humans and animals. Physicians and veterinarians have both noticed that antibiotics that were once effective at low doses must now be given in much higher doses to both humans and animals to accomplish their purpose.

The use of antibiotics at sub-therapeutic levels as a growth promotant or to reduce the incidence of disease has become controversial. Some groups oppose the practice because they fear that by destroying the sensitive bacteria, resistant bacteria are allowed to multiply, creating the potential for antibiotic-resistant bacteria to be passed on to humans. Although the use of sub-therapeutic levels of antibiotics in animals has been practiced for decades without documented problems passed on to humans working directly with them, antibiotic-resistant bacteria have become a serious problem in human hospitals. Hence, the potential and concern that a similar situation may develop among animal workers is present.

Care must be exercised to ensure that antibiotics as well as other drugs are properly used, but not overused. Mastitis and certain respiratory diseases are among the few examples in which there are no preventive vaccine substitutes for antibiotics. While management practices can minimize the occurrence of mastitis, etc., antibiotics are needed once the infection is established.

Certain antibiotic dosage forms can upset normal body functions. Some may "sterilize the gut," (kill the beneficial bacteria that aid in digestion and compete with harmful bacteria and fungi), making animals susceptible to enteric upsets and infections. Many shepherds give yogurt (which contains cultures of beneficial digestive bacteria) to a lamb after antibiotic therapy or illness to reestablish the "friendly" bacteria.

Many times antibiotics are used when they are of no benefit whatsoever, as in the case of disease caused by viruses. When the exact cause of sickness is unknown, there is a temptation to give a shot, usually a wide-spectrum antibiotic, to see if it helps. Ideally, any illness should have an accurate diagnosis to see if any antiserum or vaccination is available and what, if any, treatment by antibiotics will be effective.

BIOLOGICALS/VACCINES

Vaccines, bacterins, and toxoids are intended solely for disease protection. They have very little if any effect in treating the disease. These immunizing agents are proteins called antigens that only stimulate the sheep's immune system to produce protection against the particular disease. It should be well understood that vaccination and immunization are not the same thing, because administration of the antigen by vaccination will result in immunization only if the sheep's immune system is normal and functioning. Vaccination must be accomplished well ahead of the period in which disease exposure may occur, because it takes approximately one month for maximum immunity to develop. Very low levels of protection are observed at two to three weeks after vaccination, and it can take up to forty-five days after the last dose of some vaccines for maximum protection.

Immunizing agents fall into one of four classes, and all are commonly called vaccines:

- Bacterins, containing killed bacteria and/or fractions of the bacterial cell.
- Toxoids, containing the inactivated toxins produced by bacteria, usually clostridial organisms such as tetanus and Overeating Disease.
- Vaccines, derived from viral agents.
- Anti-serums, often called serums or antitoxins, are derived from the serum of hyperimmune animals, one that has received multiple doses of vaccine to confer a high specific antibody level against the particular disease.

Anti-serum. When an anti-serum is injected we are only "borrowing" antibodies produced in another animal to confer a temporary or "passive" immunity for a short period—usually from ten to twenty-one days. It is used to protect animals for a short time when disease is present in the herd, and to treat infected animals as an aid to overcoming disease.

In a unique situation, anti-serum may sometimes be administered along with a vaccine to give immediate protection while the animal is developing its own active immunity. Check with your veterinarian before administering serum and vaccine together, because in some instances the hyperimmune serum will neutralize the vaccine.

Bacterins. Bacterins are suspensions of bacteria that have been grown in culture media and chemically or heat killed. They are not capable of producing disease and can be used without danger of spreading disease. The bacteria used in the production of the various bacterins are highly antigenic strains isolated from animals that have succumbed to the particular disease.

Bacterins are often suggested as an aid in establishing immunity to specific diseases. Always follow the manufacturer's label. Most bacterins require a primary "priming" injection followed by a booster in one to four weeks. More often than not, very little immunity is obtained following the first priming injection. The actual protection is obtained following the booster shot. Bacterins do not confer long-lasting immunity. Maximum protection is usually for six months to a year at best between boosters.

Toxoids. Toxoids are solutions of inactivated toxins, derived from bacteria that cause disease by producing toxins that enter the blood stream and cause severe tissue or nerve damage (such as tetanus, overeating, blackleg). Since it is the toxin produced by the bacteria and not the bacteria itself that causes disease, toxoids stimulate the sheep to produce neutralizing antibodies against the toxin, thereby protecting against their deadly effect.

Vaccines. A vaccine is a modified live or killed biological preparation, which when injected into the animal (or instilled intranasally as in the case of Nasalgen), stimulates the animal's immune system to build its own protective antibodies. Modified live (MLV) vaccines contain strains of the virus that are incapable of causing the disease but still retain the immune-stimulating potential of the disease-causing strain. With few exceptions, MLV vaccines produce greater and longer lasting protection than the inactivated (killed) virus vaccines. It takes approximately two weeks for protection to appear (except for Nasalgen P13 vaccine which stimulates resistance to infection in approximately three days), and the immune response will maximize in about one month.

Store all immunizing supplies in a cool place, but do not allow them to freeze. Purchase vaccines from a reputable source, for if they are not properly stored or transported before you buy them, they may be worthless.

IMMUNIZING SHOTS

Some vaccines are applied by scratching the skin, some are sub-cu, some are intramuscular, and some are sprays into the nasal cavity. You must follow the directions of the manufacturer or veterinarian very carefully regarding both the dosage and the manner of administration. Vaccination sites on ewes and lambs would ordinarily be on the side of the breast bone (lower chest wall behind the elbow) or side of the neck. Do not inject vaccines into the armpit under any circumstances. For best results, always follow directions.

What "shots" are necessary? There is no hard and fast rule, as it will depend on what part of the country you are in, the climate, the type of operation, the prevalence of sheep flocks nearby, the purchasing of new animals, and the conditions under which the sheep must be raised.

In general, some of the most useful immunizing agents are:

- Nasalgen-IP (P13), to protect against certain types of pneumonia and respiratory diseases.
- Footvax, foot rot vaccine, if needed. This product contains an oil adjuvant (immune enhancer) that can cause injection site swellings and occasional small abscesses, so be sure to inject high up on the side of the neck.
- Covexin-8, immunizes against all the common clostridial diseases, including tetanus.
- EAE-Vibrio combination, to protect against the two common disease-caused abortions (problems more in some areas of the country than others).
- BVD vaccine (bovine) for Border Disease (Hairy Shaker Syndrome). This is also a greater problem in certain areas. Check with your veterinarian first.
- Ovine Pili Shield, the new vaccine (given to ewes) to immunize lambs, through the colostrum, against scours caused by *E. coli* bacteria.
- Selenium-vitamin E, for ewes (see Chapter on Ewes). Selenium-E is not a vaccine, but an injectable essential nutrient needed for protection of lambs and adults against White Muscle Disease and *immune deficiency*. While most sheep-rearing areas are deficient in selenium, the degree will vary according to geographical areas. There are several products on the market with different concentrations. Read directions carefully and seek professional advice, because too much can be highly toxic.

OPP
(Ovine Progressive Pneumonia)

Any disease can cause chronically thin sheep, but there is one disease called Ovine Progressive Pneumonia (OPP) that accounts for many of the persistently thin adult sheep (if nutrition and parasites have been eliminated as cause). OPP is a slow virus, similar to AIDS in humans, taking at least two years to manifest its symptoms. The virus slowly causes progressive lung damage. Ewes gradually lose stamina and body weight and have serious breathing problems, ending in physical weakness and fatal pneumonia.

While there is at present no cure and no vaccine against OPP, there are new tests that make disease control possible. To ensure that no OPP problems occur, it is necessary to have annual testing of all breeding animals (with the elimination of infected animals), and to be sure that you purchase only OPP-free breeding stock replacements. Since it is transmitted from ewe to lamb primarily through milk, any valuable breeding ewe could be isolated from the flock, and her lamb taken immediately at birth and raised on colostrum-replacer and lamb milk-replacer. This is almost 100 percent effective.

Any animals that test positive for OPP should be isolated from the rest of the flock, since transmission can result from close contact with infected animals, mainly via respiratory secretions when animals are confined to crowded quarters. Actually, all sheep that test positive will not in fact come down with the disease. However, once the signs of disease appear, the outcome is always fatal. Any animals testing "positive" should be isolated and culled, if a flock is to be protected. Any animals considered for purchase should be previously tested in order to protect your sheep from disease.

Since some local veterinarians do not really specialize in sheep, they may be unaware of OPP, and equally unaware of the testing that is available. The National Veterinary Services in Ames, Iowa will do seriological tests of blood serum appropriately drawn, the agar gel immunodiffusion test (AGITOT). At Cornell University, there is also the indirect immunofluorescent test (IIFT) and an ELIZA test for OPP. This last appears to be much more sensitive, being able to test lambs only several months old, rather than having to wait until a sheep is one or two years old to detect OPP.

OPP is another example of a "purchased disease" and certainly underlines the need to be extremely careful when buying your initial flock and any replacement animals. Request proof that the flock has been tested for OPP.

CHAPTER 17

Wool and Shearing

SHEARING

MANY COUNTY AGRICULTURAL SERVICES sponsor shearing schools in early spring for one or two days, at a nominal fee. They usually limit their instruction to electric shearing, but what you would learn there would be valuable with either electric or hand shears. We have used both and prefer the hand shears both for our own sheep and those we shear around the neighborhood. For photos and instruction on shearing with electric clippers, the manufacturers of the equipment are the best source of information.

"Rigged" blades have a leather strap taped onto the left handle (for right-handed use) and a rubber stop is taped to the top of the right handle, at the base of the blade. These hand shears are more comfortable to use, and the strap prevents them from being kicked out of your hand.

ADVANTAGES OF HAND SHEARS (BLADES)

1. Inexpensive way to get started. Order from any sheep supply catalog.
2. Need no electricity; you can shear any place.
3. Easy and quick to sharpen with just a hand stone.
4. Very lightweight, easy to carry with you.
5. Do not shave the sheep close, minimize loss of body heat in cold or rain.

SHARPENING HAND SHEARS

To sharpen, reverse the normal position of the blades, crossing over each other. Using a medium sharpening stone, follow the *existing* bevel of each blade, with long strokes. Do not sharpen the "inside" surface of either blade. If there are any slightly rough edges when you are through sharpening, run the stone flatways along the inside surface of the back (not the edge) to remove the edge burrs. For touch-up sharpening while shearing, close the shears firmly so that each cutting edge protrudes beyond the back of the other blade. Using the fine side of a small ax-stone, follow the existing bevel of each blade.

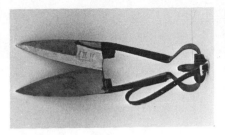

"Rigged" sheep shears.

Sharpening hand shears.

HOW TO SHEAR

The "trick" in shearing is not just the pattern of the shearing strokes, which lessens the time involved in removing the wool, but is the immobilizing of the sheep by the various "holds" that give the sheep no leverage to struggle. A helpless sheep is a very quiet sheep. This cannot be done by the use of force alone, for forcible holding will make the sheep struggle more. Try to stay relaxed while you work.

Note both the holds on the sheep, often by use of the shearer's foot or knee, and also the pattern of shearing in these illustrations.

You may like to use a *shearing belt* to lessen the strain on your back. This one is cut from a discarded inner tube and has a wide buckle. The width of the available buckle determines the width to cut the belt at both ends, and it should be wider in the middle part which fits across your back.

At one time, shearing cuts were dabbed with tar, to help them heal and to keep away flies. Shearing cuts heal quickly, but use an antibacterial spray, for they can become infected, sometimes resulting in the infection spreading to the lymph glands. While commercial shearers do not do this, a person shearing his or her own sheep will have more incentive to do it.

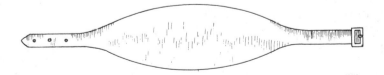

This shearing belt can be cut from an inner tube.

SHEARING IN TWENTY STEPS

Slip left thumb into sheep's mouth, back of the incisor teeth, and place other hand on sheep's right hip.

Bend sheep's head sharply over her right shoulder, and swing sheep toward you.

Lower sheep to the ground as you step back. From this position you can lower her flat on the ground, or set her up on her rump for foot trimming.

Start by shearing brisket, and up into left shoulder area. One knee behind sheep's back, other foot in front.

Sheep is on her left side. Trim top of head, then hold one ear, and shear down cheek and side of the neck as far as the shoulder, into the opening you made at the brisket.

Place sheep on her rump, resting against your legs. Shear down the shoulder while she is in this position.

With sheep in this position, and with you holding her head as shown, shear down the left side.

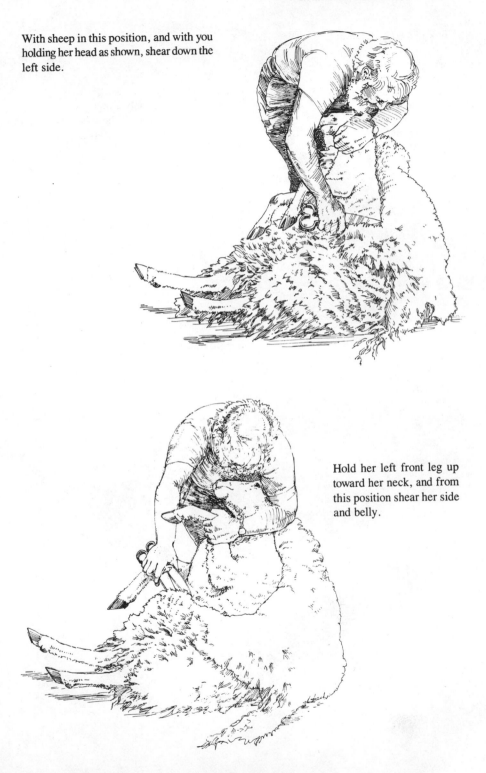

Hold her left front leg up toward her neck, and from this position shear her side and belly.

With only a minor shift in the position of the sheep, you are now ready to shear the back flank.

By pressing down on the back flank, the leg will be straightened out, making it easier to shear.

From this position, the sheep is shorne along her backbone, and a few inches beyond, if possible.

By holding up the left leg it is possible to trim the area around the crotch.

The job is half done. The shearer's feet are so close to the sheep's belly that she cannot get up.

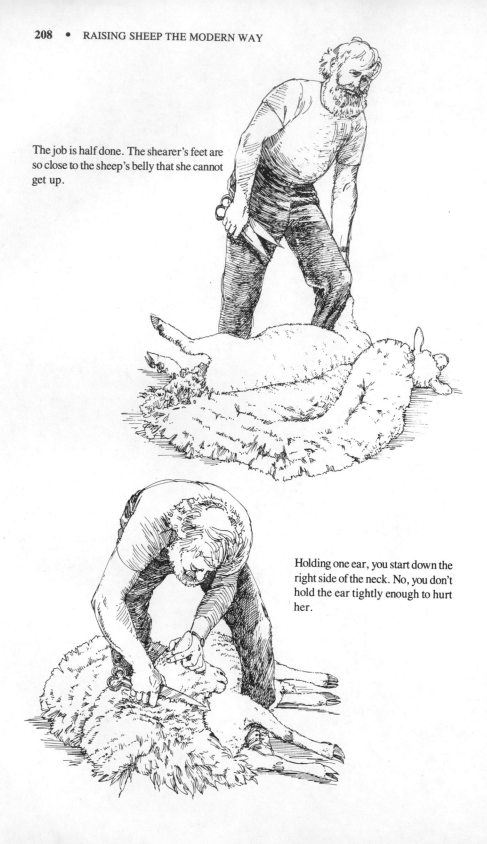

Holding one ear, you start down the right side of the neck. No, you don't hold the ear tightly enough to hurt her.

Shearer holds sheep with left hand
under her chin and around her neck,
and shears the right shoulder.

Sheep is pulled up against the shearer,
to expose her right side, so that he can
shear down that side.

Shifting position, as shown in this photo, he shears farther down the side and the rump.

Shifting his position, he finishes the right flank and shears the sheep's rear end.

He again moves his position, and, holding up the rear leg, he shears the right side of the crotch.

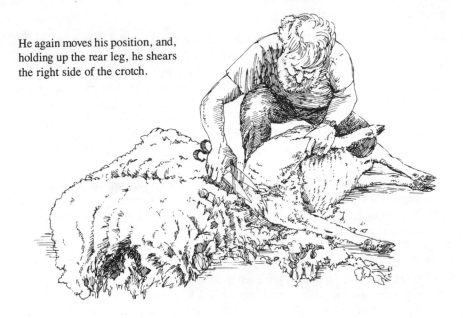

The job is done, the sheep is back on her feet, and within a minute, is eating grass again.

SHEARING SUGGESTIONS

1. Shear as early as the weather permits. Shearing nicks will then heal before fly season. Ewes can be sheared (gently) before lambing, making it easier to help the ewe if necessary, and removing dirty wool tags the lamb might suck on.
2. Never shear when the wool is wet or damp. It is very hard to dry it enough to sack or store. Damp wool is combustible, and can also mildew.
3. Pen the sheep in the afternoon prior to shearing, so they will not be full of feed when sheared. A covered "holding pen" with slatted floor is ideal.
4. Shear on a clean tarp, shaken out after each sheep, or on a wood floor that can be swept off.
5. Shear fleece in one piece, but don't trim legs or hooves onto the fleece.
6. Remove dung tags, and do not tie them in with the fleece.
7. Avoid making "second-cuts," going twice over the same place to tidy up, or overlapping your strokes.
8. Roll fleece properly, and tie with paper twine, if selling to a wool dealer or in a wool pool.
9. If selling to handspinners, pack unrolled fleece gently into empty paper feed bag, one fleece to a bag, or lay it out into a large shallow box. You can shake out much of the junk and second cuts before bagging to make the fleece more valuable.
10. "Skirting" the fleece (removing a strip about 3 inches wide from the edges of the shorn fleece) is proper, especially if selling to spinners. A slatted "skirting table" will make this easy, and will also allow any second-cuts to drop off, if fleece is thrown onto table with sheared side down.
11. Be sure you shear black sheep and white sheep separately, sweeping off the floor between each. Do not contaminate white fleece with dark snips, or vice versa.
12. For spinning wool, get top price for quality (clean fleeces without manure tags, skirtings, and vegetation).
13. Ask lower prices for lower quality fleeces, and explain the lower price to the customer. These fleeces may be quite adequate for quilt batts, rug yarn, or felting.

THE WOOL

ROLL AND TIE FLEECE

The acceptable method of rolling, in this country, is to spread out the fleece, skin side down, and fold side edges in toward the middle. Then, fold neck edge in toward the center. Last, start rolling from tail-end of fleece, and make a compact roll. Using paper twine, tie around one direction, cross the twine and tie around the other direction, and knot securely. A slip knot in the starting end of the twine makes it easy to cinch it tight. Then loop the twine around the other direction and knot it.

WOOL FOR HANDSPINNERS

If you set aside your best fleeces to sell for handspinning, be sure they are absolutely dry. They can be stored for a while in a plastic bag, but this is not best for long-term storage. With nice fleeces to sell, it seems unlikely that you will have to hold them very long before they are sold. Find where the nearest craft classes are given, and let it be known that you have fleeces to sell.

In addition to being free from seeds, burrs, stained wool, and tags, the fleece should not have "weak staple," from illness or other cause, as this would make it worthless for spinning.

Weak staple can be caused by illness, pregnancy, undernourishment, or poor nutrition, and is called "tender" wool. To test for weakness, stretch a small tuft of wool between both hands. Strum it with the index finger of one hand. A sound staple will give a faint, dull, twanging sound, and will not tear or break.

SHEEP COATS

One way of having clean fleece is to put coats (or sheep blanket or covers as they are called in supply catalogs, or "rugs" as they are called in some countries) on your sheep. Sheep coats have been tested in Australia and New Zealand, and in this country quite extensively at the University of Wyoming.

The protection of the coat not only increases quantity of *clean* wool, which is expected, but it also results in from 13 to 27 percent longer staple length of wool, and improved body weight even under harsh range conditions. It also makes shearing much easier, partly because the fleece is cleaner. In areas of severe winters, the sheep can conserve energy. This shows up in the maximum percentage of increased wool growth, and a slightly heavier birth weight for lambs born to ewes wearing covers.

One sheepman reported no death losses from coyotes during coat use, and thought perhaps the sound created by the plastic coats as sheep moved, or the sight of the different-colored coats, warded off the predators.

Cost seemed the main factor in making sheep coats less than practical. Cotton coats were not durable around barbed wire or brush pasture. Sturdy nylon-based coats were more durable, but had the disadvantage of making the sheep sweat during warm weather or close confinement.

Woven polyethylene sheep coats were found most practical during large-scale tests in Australia. Being woven, they allowed the wool to "breathe," so hot weather was no problem. Because they partially protect wool from rain, the coats minimize fleece rot and skin disease, according to Australian findings. (See Sources.)

Coats are put on the sheep right after shearing. When sheep are sheared very early, this minimizes the risks of deaths from thermal shock. Try covers on a few of your sheep, if you are wondering how they would affect the wool of your particular breed. Compare with uncovered fleeces, after a year.

The patterns shown here can be made from woven-plastic feed sacks, with heavy wide elastic used for the leg loops. This material resembles the most satisfactory of the commercial variety.

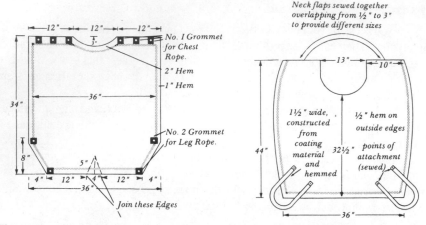

These are two styles of coats to protect sheep, with Number 10 duck or canvas used in most cases. The right pattern can be made in three sizes, with the large having a half-inch overlap on the neck flap, and 27-inch leg loops. Medium has a 1½-inch overlap and 24-inch loops, while the small has a 3-inch overlap and 24-inch loops. The loops are 1½-inch strips of the coat material, hemmed. Left pattern has grommets used for the chest and leg ropes.

When using these coats for young growing sheep of a long-wooled breed, check the fit after six or eight months, to be sure the coat is not becoming too tight. Elastic, rather than fabric, loops are better for this reason, although the elastic does have a shorter use-life and will need replacing annually.

BEAUTY OF FLEECE

Heredity determines the wool type, but its quality and strength depend on the health and nutrition of the sheep during each year of fleece growth. One serious illness can cause tender, brittle wool, a weak portion in every fiber of the whole fleece.

Contamination by vegetable matter (weed seeds, hay, straw, sawdust, or wood shavings used for bedding) can lower the value of the fleece, and even make it worthless to a handspinner, no matter how nice the wool fibers.

The beauty, luster, elasticity, and strength of the wool will suffer if the sheep's diet is deficient in protein, vitamins, and minerals. Mixed grain rations usually have protein content marked on the labels, and feed stores stock various sheep pellet rations that are convenient to use. Pasture, grain, and hay provide vitamins, and hay from sunny areas is reported to have a higher vitamin content. Vitamin supplements are available and make a very significant difference in the health of older ewes. A handful of brewer's yeast tablets (inexpensive) fed daily can extend their lifespan.

Minerals can be supplied in a variety of ways. Trace minerals can be mixed with salt or obtained in a pre-mix or as a salt block. Salt blocks seem more convenient, but are harder on sheep's teeth than loose salt. Avoid salt or mineral mixes intended for cows—they contain toxic amounts of copper for a sheep diet. You seldom know what mineral may actually be low or missing in the soil of your locality, so some type of mineral additive is good, for lack of minerals can cause specific ills, directly noticeable in the wool. Ticks (keds) can reduce wool quality, and a large number of ticks can undermine the health of the lambs and reduce wool production.

WOOL GRADING: "COUNT," "BLOOD," OR "MICRON"

Some farm flocks specialize in wool-type sheep, others primarily in meat-type, but generally the trend is toward all-purpose breeds, fairly good on both meat and wool production.

The "blood" grading and the "count" or "micron" system of wool classing do not much affect the small flock owner with no great amount of wool to sell, but the explanations may be of interest.

One method of designating the grade of wool is the "spinning count," which originally meant that one pound of fleece wool of a particular designation would spin that many "hanks" of wool, a hank being 560 yards. So, 70s would spin 70 hanks, and 60s would spin 60 hanks. The count system usually went only as fine as 80s count, but German Saxony Merino has been known to grade 90s, where one ounce of the single fibers laid end to end, would stretch 100 miles! Count is used more in foreign countries than here, and is always expressed in even numbers.

Another way is the "blood" system of grading the fineness of wool, which originally indicated what fraction of the blood of the sheep was from the Merino breed, which produced the finest diameter wool. This term no longer relates actually to Merino or part-Merino blood, but qualifies the degree of fiber diameter.

The "micron" system is an industrially accurate measurement of the average diameter of wool fiber, a micron being the number, the coarser the wool.

The accompanying chart shows the relationship of the "blood," "count," and "micron" systems, giving examples. Examples are approximate, and there is variation within most breeds.

"BLOOD" AND "COUNT" AND "MICRON" SYSTEMS COMPARED

Blood System Grades	Count System Approximations	Micron System Approximations	Examples
Fine	64s, 70s, 80s	19-20 or less	Merino Rambouillet
1/2 Blood	60s, 62s	22-24	Rambouillet Romeldale
3/8 Blood	56s, 58s	25-28	Corriedale Southdown Suffolk
1/4 Blood	48s, 50s	29-31	Oxford Dorset Romney
Low 1/4 Blood	46s	32-33	Romney Oxford
Common and Braid	36s, 40s, 44s	34-39	Leicester Lincoln Cotswold

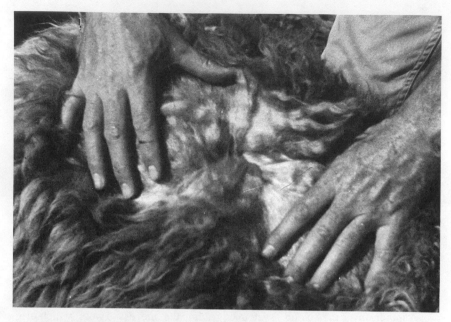

Clean silver-gray Romney fleece. This is the palest shade of black-sheep's wool.

LUMPY WOOL

One disease that can destroy the fleece is *mycotic dermatitis*, a chronic skin infection that mats the wool and makes it hard to shear, and of little value. It also predisposes the sheep to fleeceworms (maggots). A well-nourished sheep with a dense fleece is not so prone to this infection.

Treatment. Injections of penicillin-streptomycin, and a dipping or spraying with a 0.2 percent solution of zinc sulfate are useful in checking the spread of the infection.

WOOL FOLLICLE DENSITY IN VARIOUS BREEDS

Sheep Breed	Average follicles per square inch	Sheep Breed	Average follicles per square inch
Merino (fine)	36,800-56,100	Border Leicester	10,300
Merino (medium wool)	36,100-51,600	English Leicester	9,290
Merino (strong wool)	34,200-41,900	Lincoln	9,420
Polwarth	28,400-34,800	Swedish Landrace (fine)	9,350
Corriedale	14,800-19,400	Swedish Landrace (carpet)	8,260
Southdown	18,100	Cheviot	9,420
Dorset Horn	11,900	Welsh Mountain	8,900
Ryeland	10,300	Scottish Blackface	4,520
Suffolk	13,200	Wiltshire	7,350
Romney Marsh	14,200		

H. B. Carter, *Animal Breeders Abstract*, 1955.

CHAPTER 18

Meat and Muttonburger

DON'T HESITATE to put mutton in your freezer because you fear it may be tough and inedible. The leg-of-mutton makes a wonderful "smoked ham," and most of the rest can be trimmed and boned out for use as ground meat. I have assembled here a few tested "muttonburger" recipes that give make-ahead meals, casseroles, quick-fix recipes, large recipes that make one for now and several to freeze, and some sausage recipes, too. You will find that the money you can get for your culls is not nearly as much as they are worth when you put them in your freezer.

To assure more tenderness of either lamb or mutton, have your slaughterhouse use the "Tenderstretch" hanging method.

TENDERSTRETCH

Texas A-&-M University developed a method of carcass hanging which improves the tenderness of most of the larger and important muscles of the loin and round (most of the steaks and roasts). It is called "Tenderstretch," and consists of suspending the carcass from the aitchbone, within an hour of slaughter. The trolley hook should be sterilized before inserting in the aitchbone on the kill floor.

This method does not require any change in equipment in small slaughterhouses, and is also suitable for farm use.

It prevents the shortening of the muscle fibers as the carcass passes into rigor mortis. Before that, the muscles are soft and pliable, and if cooked rapidly are very tender. But after rigor mortis has developed, the shortened muscles become fixed and rigid. It takes an aging period of seven to fourteen days at temperatures between twenty-eight and thirty-four degrees before the muscles lose their rigidity and become more pliable.

With Tenderstretch hanging right after slaughter, meat is as tender in twenty-four hours of chilling as if it were aged, and a further aging will improve the tenderness even more. Thus, with little extra effort, and no additional cost, there is a greater improvement in the tenderness of many of the important cuts in the animal. It does not produce the mushy overtenderness that sometimes results with enzyme-tenderized meat.

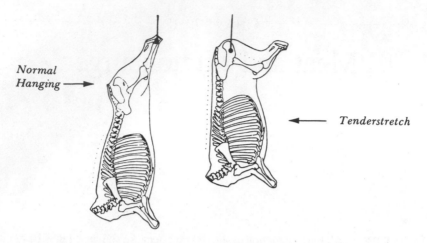

This shows the normal (left) and the Texas A & M "Tenderstretch" method (right) of hanging of a carcass. Points of hanging are the Achilles tendon, in left illustration, and the *obturator foramen* of the aitch bone, at right.

CUTTING INSTRUCTIONS FOR MUTTON

To get the maximum use and enjoyment of a mutton carcass, give these instructions:

1. Cut off lower part of hind legs for soup bones.
2. Have both hind legs smoked, for leg-of-mutton "hams."
3. Package riblets (spareribs) and breast in two-pound packages, to pressure-cook and separate meat, for curry recipes. These parts are hard to bone out.
4. Have tenderloin removed and made into boneless cutlets.
5. Have the rest boned out, fat trimmed off, and ground. Double wrap in one-pound packages and try my muttonburger recipes in this chapter. Do not be surprised when the ground mutton seems a lot juicier than in other ground meat. Older animals' tissue can bind large amounts of water.

CUTTING PLAN FOR LAMB

This slightly different plan of cutting a lamb is simple, and not far from the standard cutting procedure. It was the suggestion of Mrs. Nancy Wetherbee, and was shown in *Shepherd* magazine in 1968.

The rib on the chops is cut very short. This leaves no uneaten waste, makes a much nicer chop and longer spareribs.

They can be braised, or pressure-cooked and all the lean meat separated from the bone and fat, for a delicious curry.

With this type of cutting, the larger the lamb the better, if young and not overly fat. If the leg is too large, remove a few more chops and steaks, either from the center of the leg or from the loin end.

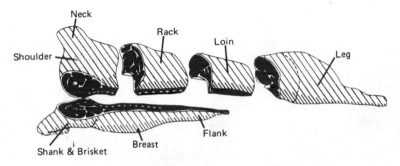

There are several correct ways to break a lamb carcass, and no one method can be considered best. However, for many purposes, the method shown is ideal. (*Lamb Cutting Manual*, published by American Lamb Council and National Livestock and Meat Board)

A STANDARD CUTTING DIAGRAM

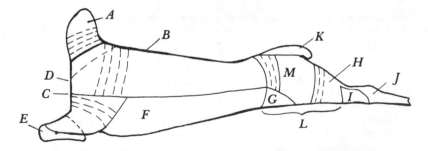

Here's a detailed method of cutting a carcass, suggesting uses for the individual cuts. (A) Neck: slice for braising or bone for stew or ground lamb. (B) Loin, rack, shoulder: cut for roasts or cut between ribs for chops. (C) The lateral cutting of the carcass is at a high level, leaving less rib ends on the chops and more in the spare ribs. (D) Lift shoulder blade for roast; bone if desired. (E) Braise trotters or cut for soup. (F) Breast: cut between ribs for braising. (G) Grind or cut for stew. (H) Center-cut steaks. (I) Braise trotters. (J) Discard or use for soup. (K) Discard or use for soup. (L) Leg. (M) Optional removal of sirloin steaks from leg roast. (*Shepherd* magazine)

RECIPES

Smoked Leg of Mutton "Ham"

Glaze: 1/2 cup brown sugar, firmly packed
1 teaspoon prepared mustard
1 cup orange or pineapple juice
Cloves

Method: Soak mutton in cold water for one hour. Dry with paper towels and wrap securely in a large piece of aluminum foil. Seal edges well and place in a baking dish. Bake at 350° for 30 minutes to the pound.

Mix together brown sugar, mustard, and fruit juice. Place the precooked leg in a baking dish. Score outer covering with knife and pour juice mixture over it. Stud with cloves. Bake the leg for an additional 40 minutes, basting often with pan juices. Serve hot or cold.

Note: Simmering may be preferred for the first stage. Soak mutton as above. Plunge into large pan with warm water. Bring to a boil and simmer for 30 minutes per pound, or until tender. Allow leg to cool in the liquid. Drain and refrigerate, covered (do not freeze) until needed, then bake with glaze.

In our experience, the yearling or two-year-old cut is very tender and tasty. But the old, old, ewe is tasty and not very tender. So, cut the real old ham into several pieces that will fit into your pressure cooker. Do the one-hour soak, then pressure-cook for 15 minutes at 15 pounds of pressure. I then bake the "ham." It will not need a long baking and will be both tasty and tender. Grind up leftovers for ham hash. Cook split peas with the bone.

Mutton ham recipe is from the Australian Meat Board, as printed in *Shepherd* magazine in 1973.

Mutton Patties for the Freezer

5 pounds lean ground mutton	1 teaspoon crushed mint leaves
5 tablespoons water	1 teaspoon fenugreek (optional)
4 tablespoons lamb seasoning salt	1 beaten egg
1 teaspoon coarsely ground black pepper	1 tablespoon wheat germ

Mix all together. Press into patties with hamburger press (or divide in about 1/4 pound balls and press flat). Package for freezer in desired number of servings to the package, separated by a double thickness of wax paper or foil. Double wrap in plastic or freezer paper.

To serve, thaw for 45 minutes in the package and then separate, or separate and thaw for 25 minutes.

Broil, pan fry, or bake in gravy.

Seasoning Salt for Lamb and Mutton

1 cup fine plain popcorn salt (dissolves instantly, doesn't fall off the meat)
1 teaspoon freshly ground black pepper
1 teaspoon paprika
1/2 teaspoon ground ginger

1/2 teaspoon dry mustard
1/2 teaspoon poultry seasoning
1/3 teaspoon cayenne pepper
3 teaspoons garlic salt (or 2 teaspoons garlic powder)

Combine all ingredients. Mix well and pack in shaker jars.

This recipe originated at Purdue University's Home Economics Department, and was reprinted in *Shepherd* magazine.

Breakfast Sausage

1 pound lean ground lamb or mutton
1/8 teaspoon coarsely ground black pepper
1/2 teaspoon salt (or more)

1/4 teaspoon powdered marjoram
1/4 teaspoon powdered thyme
1/4 teaspoon powdered sage (or more)

1/4 teaspoon savory

Mix all ingredients together thoroughly. Cover bowl, place in refrigerator overnight. To use, shape into patties about 1/2 inch thick. Cook over moderate heat in heavy skillet until brown. Turn. Brown other side, lower heat to cook through. Serves 5-6.

If you like your sausage a little more moist, you can add about 2 tablespoons water and cover the skillet, when you lower the heat to cook. For a larger quantity to freeze in rolls for slicing later, add a little ice water and mix in with sausage, so it doesn't crumble when you defrost and cut it into slices.

This recipe was in *Shepherd* magazine, April, 1972.

Lamburger Helpmate

1/2 cup all-purpose flour
2 tablespoons grated Parmesan cheese
4 beef bouillon cubes, crushed, or 4 teaspoons instant bouillon

1/2 teaspoon garlic powder
1/4 cup dehydrated onion powder, or dried onion flakes
4 tablespoons nonfat dry milk powder

1 teaspoon thyme

Mix all together and store in tightly covered jar.

To use: Add about 1/2 cup of the above mixture to one pound browned ground lamb or mutton (drained of all fat). Add tomato juice or water and your choice of uncooked macaroni, noodles, or rice. Simmer, covered, until pasta is tender, adding more liquid as necessary.

Three cups of liquid, plus "helpmate" mixture and one pound of meat, would suffice for about 1 cup of macaroni, 1 1/3 cups of noodles, or 3/4 cup of rice.

1. Square up the lamb breast section.

2. Lift the top cover.

3. Stuff with lean ground lamb.

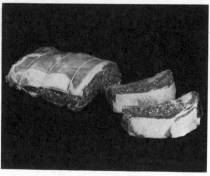

4. Economy Scotch roast and chops.

Lamb Scotch Roast and Chops

The lamb Scotch roast and chops are both excellent ways to use the lamb breast section and lean trimmings all at one time. Square off the breast by removing the brisket point and flank section. Run knife along the top of the rib bones and lift the top cover. Leave the top cover attached at the brisket side. Stuff the breast section with lean ground lamb, and either tie or net the Scotch roast.

For Scotch chops, let the cut firm up in the cooler, then cut in between each rib. (National Livestock and Meat Board)

Sloppy Joes

1 pound ground mutton or lamb
1 tablespoon dried onion flakes, or 1/2
  small onion, chopped
1 teaspoon garlic salt
1/4 teaspoon curry powder
1/4 teaspoon ginger
1/2 teaspoon coarsely ground black pepper
1 can tomato paste (6 ounces)

1/2 cup water
2 tablespoons brown sugar
3 tablespoons lemon juice
1 tablespoon soy sauce
2 teaspoons chopped dried or fresh
  parsley
1 teaspoon Worcestershire sauce

8 sliced burger buns

Brown ground meat and onions, pour off any extra fat. Season. Combine tomato paste and rest of ingredients, except buns, and add to meat mixture. Bring to a boil. Serve meat mixture on toasted buns, 1/4 cup of mixture to each bun. Serves 8.

Vi's Tamale Pie

Meat mixture is sufficient for 2 tamale pies; freeze half for later use

3 pounds ground mutton or lamb
1 green pepper, chopped
1 large onion, chopped
1 15-ounce can tomatoes
1 15-ounce can tomato sauce
1 6-ounce can tomato paste (optional)

1 cup sliced ripe olives (or more)
2 teaspoons sugar
1 teaspoon salt, or to taste
1 tablespoon *each* chili powder and
  powdered cumin
Cayenne pepper (optional)

Saute meat, green pepper, and onions together until meat loses its red color. Add tomatoes, tomato sauce, tomato paste, olives, sugar, and salt to taste. Add chili powder and cumin seasoning. Simmer 15 minutes, and then taste for seasoning. Add more chili powder if desired. Add some cayenne pepper if you want it hotter. Divide meat mixture, freezing half for later use.

Cornmeal Crust

1 cup yellow cornmeal
1/2 cup shredded cheddar cheese
1 teaspoon salt

Mix cornmeal with 1 cup cold water, then stir it into 1 cup boiling water. Cook slowly, stirring constantly, until thick. (If cooking in microwave, stir every 30 seconds until thick.) Spread cornmeal in shallow baking dish, reserving 1/2 cup to decorate the top. Spread meat mixture over cornmeal. Decorate edge with small spoons of cornmeal. Bake 30 minutes at 350°. Spread shredded cheese on top, bake 5 minutes more. Let stand for 10 minutes before serving. Serves 6.

Easy Meat Pie

1 pound lean ground lamb or mutton	1/4 teaspoon garlic powder
1 large onion, chopped	1 10-3/4 ounce can cream of potato soup
2 beef bouillon cubes	Salt to taste
2 tablespoons boiling water	Pastry for top and bottom crust

Saute ground meat and onion slowly in skillet, just until meat barely loses its red color, separating with a fork. Melt bouillon cubes in boiling water, add to meat, and mix in garlic powder and undiluted soup. Salt, to taste. Pour into pastry lined 9-inch pie tin, cover with top crust. Crimp edges to seal, and slit the top crust in several places. Bake 15 minutes at 425°, reduce heat to 350°, and bake 45-55 minutes longer. If top is getting too brown, cover loosely with aluminum foil. Serves 6.

Garden Meat Loaf Squares

2 tablespoons vegetable oil	1 cup bread crumbs
2/3 cup chopped onions	2 teaspoons salt
1 cup fresh string beans, cut small, or	1/2 teaspoon freshly ground black pepper
drained canned beans	1 tablespoon Worcestershire sauce
1/2 cup green pepper, chopped	1 teaspoon soy sauce
1 cup celery, chopped	1 egg, beaten lightly
2 pounds ground lamb or mutton	2/3 cup tomato juice

Garnish, catsup or chili sauce

Saute onions, beans, pepper, and celery in oil until tender. Mix meat, bread crumbs, seasonings, egg, and tomato juice. Mix in the vegetables. Press mixture into a 9 x 13 x 2 inch pan. Bake 30 minutes at 350°. Spread top with thin layer of catsup or chili sauce, bake 5 minutes more. Cut into squares to serve.

Serves 8. (Or serve 4, and reheat the rest for another meal. Good reheated.)

Lamb Rollettes with Curry Sauce

1 pound ground lean lamb or mutton	3/4 teaspoon salt
1/3 cup finely shredded carrot	1/4 teaspoon ground allspice
1/2 cup fine bread crumbs	1/8 teaspoon coarsely ground black pepper
1/4 cup milk	4 slices bacon, slightly cooked, drained
1 tablespoon minced onion	Curry sauce

Lightly mix lamb, carrots, bread crumbs, milk, onion, salt, and pepper. Form into 4 small loaves. Wrap a strip of bacon around each loaf, tucking ends under loaf. Place in shallow baking dish. Bake at 350° for 25-30 minutes, or until slightly browned. Serve hot with curry sauce.

Curry Sauce: Melt 2 tablespoons butter in saucepan. Stir in 2 tablespoons flour, 1/2 to 1 teaspoon curry powder and 1/2 teaspoon salt. Gradually stir in 1-1/3 cups milk and cook, stirring, over medium heat until thickened and smooth. Add 1/2 cup cooked green peas. Heat through, stirring constantly, serve as sauce over lamb loaves. Serves 4.

This recipe originated with the American Lamb Council, and was reprinted in *Shepherd* magazine.

Anna's Casserole

1-1/4 pounds lean ground mutton
1 onion, chopped, or 2 tablespoons dried
   onion flakes, divided
1 teaspoon lamb seasoning salt, divided
1 10-ounce package frozen peas, defrosted
2 cups celery, thinly and diagonally sliced

1/2 teaspoon freshly ground black pepper
1 10-3/4 ounce can cream of chicken or
   cream of mushroom soup
1 7/8-ounce package crushed barbecue
   potato chips
Paprika

Saute ground mutton and half of the onion in skillet until lightly browned, seasoning with half of the lamb salt, breaking apart with a fork as it cooks. Drain off fat. Spoon meat into medium-size loaf pan. Scatter defrosted peas over the meat. Layer the celery on top of the peas. Mix pepper and the rest of the onion and seasoning salt with the can of soup, spread on top. Put crushed potato chips on it, and sprinkle well with paprika. Bake about 30 minutes at 375°. Serves 4.

Muttonburger Stroganoff

1 pound ground lamb or mutton
1 tablespoon butter or margarine
2 tablespoons dried minced onion
1/2 pound fresh mushrooms, sliced
1 10-1/2 ounce can cream of chicken soup

Salt to taste
Pinch of nutmeg
1 cup sour cream
Egg noodles
Freshly ground black pepper

Brown the ground meat in margarine until red color disappears. Add minced onion and sliced mushrooms and saute until tender. Add chicken soup, salt, and nutmeg. Heat to simmering.

Cook noodles until tender, and drain well. Add sour cream and noodles to meat mixture, stir gently. Reheat very hot, but do not boil. Serve on heated platter, topped with freshly ground pepper. Serves 4.

Tortilla Pie

1-1/2 pounds ground lamb or mutton
1/2 cup chopped onion
1 medium green (or Anaheim) pepper,
   chopped
1 15-ounce can tomato sauce

1 teaspoon coarsely ground black pepper
1/2 teaspoon salt, or to taste
1-1/2 cups crushed corn chips
1 cup shredded medium cheddar cheese
6 thin slices cheddar cheese

Saute meat, onions, and green pepper until meat has lost its red color, separating with a fork as it cooks. Pour off fat. Add tomato sauce, pepper, and salt to taste. In shallow casserole, put 1/3 of the corn chip crumbs, then half of the meat mixture and half of the shredded cheese. Repeat layers, then top with the rest of the corn chip crumbs. Bake 30 minutes at 350°, then top with the sliced cheese and bake about 5 minutes more. Good served with lettuce salad. Serves 4-6.

Top-of-the-Stove Meat Loaf

1-1/2 pounds lean ground lamb or mutton
1-1/4 cups quick-cooking oats
1 cup ground and drained green tomatoes
   (reserve juice)
1 teaspoon salt
1/4 teaspoon garlic powder
1/4 teaspoon freshly ground black pepper

1 small onion, finely chopped
1 tablespoon chopped parsley
1 egg, slightly beaten
1 cup reserved green tomato juice, plus
   enough red tomato juice to make the
   cup
2 tablespoons vegetable oil (less if you use
   Teflon or T-Fal skillet)

Sauteed onion rings

Combine all ingredients except oil and onion rings. Shape into a large rounded patty, slightly smaller than large skillet. Pour oil into hot skillet, place meat patty in hot oil, and cover tightly. Cook over medium heat for 15 minutes. Loosen with spatula and turn patty over, browned side up. Cut into 6 wedges, separating them slightly. Cover and cook 15 minutes, or until done. Serve topped with sauteed onion rings. Serves 6.

Recipe from *The Green Tomato Cookbook* by Paula Simmons, published by Pacific Search.

Meat Balls In Spaghetti Sauce

1 pound ground lamb or mutton
1 cup grated cucumber (or zucchini)
1/2 cup dry bread crumbs
1 egg, beaten
1 teaspoon salt

Dash cayenne pepper
1/2 teaspoon Tabasco sauce
1 tablespoon dried onion flakes
1 10-1/4-ounce can spaghetti sauce (meat-
   less)

Parmesan cheese

Combine all ingredients except spaghetti sauce. Form into small balls, cook over low heat until browned on all sides. Add spaghetti sauce. Cook over low heat about 20 minutes, serve over spaghetti, with Parmesan cheese. Serves 4.

Recipe from *Shepherd* magazine, June, 1966.

Hasty Hash

1 pound ground lamb or mutton
1 tablespoon vegetable oil
1 small onion, chopped
1/2 teaspoon salt

1/8 teaspoon garlic powder
1/2 teaspoon freshly ground black pepper
4 tablespoons soy sauce
2 cups raw potato, shredded (can be
   defrosted frozen hashbrowns)

Saute meat with oil until pink color leaves; add onions and saute until onions are transparent. Separate meat with a fork, as it cooks. Stir in salt, garlic powder, pepper, and soy sauce. Mix. Layer potatoes on top of meat, cover pan, and cook on medium-low heat for 20 minutes, stirring gently from time to time. Uncover and turn heat up a little. Stir and cook until potatoes are beginning to get brown. Good with catsup. Serves 4.

Oregon Lamb or Mutton

3 tablespoons all-purpose flour
1 teaspoon dry mustard
1 teaspoon salt
1/2 teaspoon coarsely ground black pepper
4 lamb shanks, split *or*
    3 to 4 pounds lamb neck, sliced *or*
    2 to 3 pounds lamb steaks, chops, *or*
    shoulder roast
Vegetable oil for browning

1 10-1/2-ounce can consomme
1 10-1/2-ounce can cream of mushroom
    soup
1 tablespoon Worcestershire sauce
1 tablespoon Kitchen Bouquet
1/8 teaspoon garlic powder
1/2 teaspoon curry powder
1/2 cup white wine
4 to 6 servings rice, noodles, or potatoes,
    cooked

Put flour, mustard, salt, and pepper in a paper bag. Add lamb and shake to coat. Brown lamb in hot oil in nonstick skillet and place in slow-cooking pot. Combine all remaining ingredients except rice, and add to pot; sauce should cover about 3/4 of the meat. Cook 1 hour on high, then cook on low for 6 to 8 hours, or until lamb is extremely well done. Serve at once, or pour off all liquid and chill it until the fat can be skimmed off. Pour remaining liquid back over meat and reheat. Serve on rice. Serves 4-6.

Recipe from *My Secret Cookbook* by Paula Simmons, Pacific Search Press, 1979.

Lamb Curry (or Mutton)

2 pounds lamb ribs or lamb breast
1 cup water
1 large onion, chopped
1 tablespoon curry powder, or to taste
1 cup reserved lamb liquid or water

1 teaspoon salt
1 teaspoon ground fenugreek or ground
    cumin
1 tablespoon raisins
1 apple, peeled and chopped

4 to 6 servings rice, cooked

Side dishes: chutney, ground coconut, raisins, chopped nuts or sunflower seeds, chopped green onions, chopped hard-cooked eggs, chopped green pepper, sauteed banana slices.

Pressure-cook lamb with water for 20 minutes at 15 pounds pressure; cool to reduce pressure. Drain off liquid and chill, then skim off fat. Let meat cool until it can be handled comfortably, then separate lean pieces from fat, bone, and skin. Place lean tidbits in saucepan with onion and curry powder, plus 1 tablespoon of lamb liquid; saute until onion is transparent. Add remaining lamb liquid, plus salt, fenugreek, raisins, and apple; simmer 20 minutes. Set aside at room temperature for 1 hour, then reheat and serve over rice, accompanied by side dishes. Serves 4.

Recipe from *My Secret Cookbook* by Paula Simmons, Pacific Search Press, 1979.

CHAPTER 19

Profit From Sheep

THE KEY TO PROFIT is to make good use of all your potential sources of income, connected with your sheep. This requires good planning and good management, so that all your ewes will lamb, with a high percentage of twins, and low percentage of lamb deaths. Cull out your poor producers, and replace them by keeping the best of your early-born, fast-growing twin ewe lambs.

Having heavy fleeces on your ewes and your ram will give added pounds of wool to sell. Doing your own shearing keeps expenses down. Keep your wool in good condition, and market it at the best price per pound.

Control parasites, so you are feeding sheep, not worms, and so the sheep are in good condition, which makes them more resistant to disease.

Publicize whatever superiority your breed has, to make money by selling breeding stock.

For both profit and pleasure, make use of all the by-products that you can.

POTENTIAL SOURCES OF SHEEP RELATED INCOME

1. Locker lambs
2. Mutton
3. Ram rental
4. Breeding stock
5. Pelts and pelt products
6. Shearing for hire
7. Cheese from sheep milk
8. Manure for gardens
9. Soap and candles
10. Special uses of wool
11. Building sheep "furniture"
12. Locker hooking with fleece
13. Feltmaking with fleece
14. Special breeds and colors
15. Selling wool to spinners
16. Handspun yarn and products
17. Cottage-industry processing
18. Livestock dog breeding
19. Incentive payments
20. Merchandising products

21. Teaching

LOCKER LAMBS

If you are raising only a few sheep, you will find that the meat lambs can be sold at a much better price by dealing directly with the customer. At a better price than you would get from a "lamb pool," you can still be giving them a better buy than they would get at the meat market.

Most states have some restrictions on slaughter-and-sell practices, some designed to deter rustling, others to enforce sanitation. In some states, if a lamb is taken to the slaughterhouse for inspection and butchering, and cut and wrapped, each package of frozen meat must be stamped "not for sale." However, this restriction on private slaughterhouses need not stop you from legally selling locker lambs.

Sell to your customer in advance, deliver the lamb to the slaughterhouse and give them your customer's name. They will notify you, not the customer, of the cutting weight as you direct them. Collect the price per pound on that weight from your customer, who then picks up the meat from the slaughterhouse, all cut and wrapped and frozen, and pays them for the cut-and-wrap charges. Normally, the seller of the lamb pays for the slaughter charge, which is a nominal flat fee per lamb.

Taking orders in advance is always a good idea, not just to stay legal while selling locker meat, but so you have your whole crop sold, and can deliver it about the time the summer pasture starts to dry up. Fast growth of your lambs will assure that they are ready for marketing by then.

Fast growth is also associated with tenderness, so if a lamb takes longer getting to locker size, it may not be quite as tender as if it had a rapid growth rate.

Young lamb is naturally expected to be tender but several factors can, one at a time or combined, work against this tenderness:

1. Stress imposed on animals prior to slaughter, such as rough handling when catching and loading.
2. Failure to age the carcass long enough in the chill room. One week is not too long, when hung by the standard method.*
3. Slow growth rate. This is a good reason to grain your lambs in a creep feeder.
4. Drying out in slow freezing. Most cut-and-wrap facilities will do the freezing, and faster than it could be done in your home freezer.
5. Length of time in freezer storage. One year should be the maximum storage.

Organic lambs. With the current trend to health consciousness, there is a very special market for organically-grown locker lamb, which can be met only by the small grower. Big feeders and producers have more of a disease problem than is ordinarily present on a small farm, so they have to use medicated feed as a preventive measure, even when disease is not present. (Vaccines, being immunizing agents, would not detract from the "organic" description.)

Very strict organic raising is difficult. You must provide unmedicated feeds, which means whole and crushed grain, not pelleted feeds that contain additives. If you buy your grain direct from its grower, you will know whether it has been sprayed

*For added tenderness by a new hanging procedure, see "Tenderstretch" described in Chapter 18 on Muttonburger.

or treated in any way. In this matter, a small grain producer is less apt to use a lot of chemicals, mainly because their use is only profitable on a large scale.

Extremely good sanitation practices can make medications less necessary. You can avoid chemical dipping for ticks if you once get rid of ticks completely, and do not allow them to be brought in by a strange sheep.* We loan out our rams for breeding services and have to de-tick them when they come home, or have ticks spread to all the ewes.

You should worm your *ewes* regularly, and then keep the lambs in clean surroundings with creep feed and "advance creep" pasture feeding, where they get into each fresh pasture ahead of the ewes. When this is done, there is a very good chance that you will be successful in preventing a worm build-up in the lambs that would necessitate worming them before locker age.

In general, lamb consumers are among the higher-income groups of the population, so this should determine where to advertise, if advertising is needed to sell

Owner of a yarn shop advertises her locker lambs to the customers who buy yarn and fleeces

---

* Although not officially "approved" for sheep, rotenone is one organic dip that is the choice of organic growers when they find dipping necessary. See Chapter 14 on External Parasites.

your locker lambs. Do keep in mind the consumer preferences for leaner meat, and do not overfatten the lambs in the last month prior to slaughter.

Lambs for Easter. Creep feed your lambs and try to have some of them ready for sale by Easter. The eating of lamb is part of the religious festivities in the Greek Orthodox tradition, among others. If you have lambs born early (first half of January or before) and do not have them promised, you might tell the nearest Greek Orthodox church of their availability, or advertise if there is a Greek newspaper in your area. The size preferred in the Northeast is about 35 to 40 pounds liveweight; in the West the ideal size is a little larger. Lambs sold at that size are called "milk-fed." The term "hot-house lamb" is sometimes applied to the early January lambs that are sold at Easter, and sometimes to the fall lambs, born out-of-season and raised mostly indoors, for early spring sale.

Walk-in refrigerator. Utah State College engineers, in Logan, Utah, have plans for a large walk-in refrigerator that you can build (see Sources). This would make on-the-farm processing a practical alternative to the slaughterhouse, and could also be used for vegetables from a home garden. The four-page blueprints cost about $4, and are a complete construction guide.

MUTTON

Selling an aging ewe or an extra ram is not as easy as selling a lamb, which is expected to be more delicate and tender, for mutton has a rather bad image in this country. Many people (even people who have never tasted it) say they don't like it, and expect it to be tough and strong in taste. Prejudices are hard to overcome, so consider saving the mutton for your own locker. You will be pleasantly surprised to find that there are a lot of uses for mutton, so that you will be able to utilize whatever "culls" you have and enjoy doing it.

Keep in mind the known digestibility of mutton, which makes it a good meat for people who have various digestive difficulties.

A grain-fed farm ewe or ram makes good eating. For the person who thinks he or she doesn't like it, or expects it to be tough, see the previous chapter of recipes.

The strong "sheep taste" of Australian mutton has made it difficult for their industry to sell it overseas. Now there is a new technique to make the mutton taste like lamb. By feeding a special diet based on cotton seed and sunflower seed, the flavor and cooking smell can be altered, so that a seven-year-old sheep fed on the diet for forty days is said to be able to pass very easily for lamb meat.

An old ram is bound to taste stronger than a ewe, so some producers are turning the retired rams into dog food. For feeding to their own guard dog, they use it ground and completely cooked, while selling it raw (and ground) to other dog owners. With a secondhand commercial grinder and adequate refrigeration space, they can take special orders from dog owners for a specified poundage per week.

Another use for the meat of rams or older ewes is to have it made into sausage. Ask your meatcutter or locker owner about custom sausage. Also see Sources for home sausage-making supplies.

RAM RENTAL

Providing breeding services to people who have just a few sheep for "lawn-mowers," and do not want to keep their own ram, is a little business all in itself. Ordinarily, the neighbor with four or six sheep will not want to be bothered to have his or her own ram and would much rather pay for the use of yours. The rental, for money or for a choice of one of the lambs, can help pay the keep of an "extra" ram that you might want to keep for yourself, to give you more options in breeding.

PUREBRED OR SPECIALTY BREEDING STOCK

In raising purebred and/or registered sheep on a small scale, some profits should be made in the sale of breeding stock. If they are not, you have lost some of the advantage of paying for purebred and registered sheep, for the actual receipts for sale of wool and meat would be very little different with less expensive breeding stock. To raise registered sheep successfully involves a tremendous amount of record-keeping and either experience with sheep or good planning in order to improve the flock or just to keep it from deteriorating. Beginners are often advised to start with less expensive sheep, so there is less money involved in losing a sheep due to inexperience.

After a year or two of raising no-breed-name sheep, it will be much easier to decide on the breed that will actually offer the most potential for profit, given your particular interests. Knowing how much time you can spend with them will help decide whether the most prolific breeds, which require more attention at lambing time, would be suitable for your situation.

If you are buying purebreds and plan to sell them, try to select a breed that would appeal to a market that you are familiar with, if possible, as well as one suited to your area. Some unusual breeds are in great demand for non-commercial raising, with good sales for breeding stock. Some breeds thrive at high altitudes, some do well in heat, and some prefer cooler climates. Some graze well on rolling hills, and some are more at home on flat meadows. Some breeds can tolerate abundant rainfall; others would suffer with hoof problems and fleece rot if there was too much rain.

See the section on raising black sheep, later in this chapter, if natural dark colors sound more interesting than white.

PELTS

The pelts of meat lambs can be another source of income. If you are a spinner and want both fleece and the pelt, you may prefer to shear the lamb and then wait about six weeks before it is slaughtered. This way you will have enough wool on the skin to use as a "shearling" for slippers or jacket lining. Often you can sell the products of one pelt for a much higher price than you could ask for the pelt.

For tanning, skins should not be damaged by ticks, which is another reason to keep your sheep free of them. The dark lumps caused by tick bites are called "cockle" in pelts or leather. If you are using tanned shearling pelts to make jackets with the wool inside, the outer surface can be sanded to produce a beautiful suede finish. However, cockle defects would seriously impair the softness and appearance of the suede leather.

Shearing nicks will also show up in the pelt. Skinning should be done carefully, to avoid cuts into the hide. Tanneries usually mention that if there are more than two cuts in the middle of a skin, it should be discarded.

Most slaughterhouses realize a small income from the sale of pelts they get from the animals being processed, but you can ask for the pelt back. Pick it up the day it is skinned, if possible, or the next morning.

Pelts should be liberally salted on the flesh side, as soon as possible after the body heat has left them, to avoid spoilage. If you are going to start home tanning immediately, spoilage is no problem, but if you are sending them away for tanning, salt the hides, using five pounds or more of granular salt on big skins, two pounds or more on lambs. Salt draws the moisture out of the skin. With several hides, salt each, then stack them leather side up and raised off the floor on boards. In three or four days, they will be ready to ship to a tannery. With tanning prices known in advance, you can just pack the pelts salted and folded, in a feed bag in a carton. Attach a note with your return address and phone number, and indicate whether you want natural or washable tanning (washable tanning costs more).

The machine-washable pelts (glutaraldehyde tanning) are popular as bedpads for invalids, since the pelts distribute pressure evenly, dissipate moisture, do not wrinkle or chafe, and prevent ulcers and bedsores. They are also marvelous for babies.

You can get catalog supplies to do your own tanning. To decide about the price of home tanning versus tannery tanning, estimate the cost of your materials, and also the value of your time if you have little to spare. Weigh this against the cost of postage and tannery fees.

When trying any tanning process for the first time, be cautious and do only one pelt. When you have done it once, you may see ways to do a better job than you did the first time, or may prefer to try one of the other processes to see if it is easier and more satisfactory.

Once you have perfected your system of tanning and have done it a few times, you should find a ready market in local craft shops or decorator shops. To get a better price, you can sell directly to your customers, or design and produce wearables or furnishings from the tanned pelts. While the tanning chemical is dangerous to handle and must be used with care, the results can be worth the trouble.

The salt-acid process that I will describe is not washable, but it makes a pelt that can be cleaned with solvent-and-sawdust. The acid must be handled carefully and neutralized well so that it does not remain on the skin and damage it.

Fleshing out the pelt. First, scrape the flesh side with a heavy knife to remove all meat, tissue, and grease. Do not injure the true skin, or expose the hair roots. Scrape off all tough membranes and inner muscular fleshy coat.

Salting the pelt. If you are not going to tan the skin the day it comes off the sheep, you should salt it heavily to preserve it for later tanning. As soon as the animal heat has left the pelt, rub common pickling salt into the flesh side. A lamb pelt will take about three pounds. Do a thorough job, giving attention to salting the edges well. Spread the pelt out to dry, flesh side up.

Preparing salted pelt for tanning. Later, when you want to prepare this pelt for tanning, soak it overnight in a large tub of cold water containing one cup of laundry detergent and one cup of pine-oil type disinfectant. In the morning, remove this water by spinning the pelt in the "spin" cycle of your washer. Then wash the pelt in the washing machine, short cycle with cool water and laundry detergent. Rinse. Spin the rinse water out in the "spin" cycle, then proceed with tanning.

Preparing a fresh pelt for tanning. First, flesh out the skin as directed above. Then sprinkle the skin side of the pelt with strong detergent, and brush it with a stiff brush.

Wash the pelt in the washing machine, short cycle with cool or lukewarm water and detergent. Rinse. Spin out the water, using the "spin" cycle of the washing machine, and proceed with your choice of tanning processes. All the fat, blood, and dirt should be removed from the pelt by now.

Salt-acid tanning process. For the salt-acid tanning solution, use a plastic drum or plastic garbage can. Metal containers must not be used. For best results, solution must remain at about room temperature—between 65 and 75 degrees.

Solution: For each one gallon of clear 70-degree water, use one pound pickling and canning salt and one of the following acids: 1 ounce concentrated sulphuric acid, or 4 ounce new battery fluid (acid), or 1/2 cup sodium bisulfate, dry crystals, or 2 ounce oxalic acid crystals.

Use your choice of only *one* of the above acids, with the water and salt, for tanning. A choice of acids is given so that you can use the one most easily obtained in your area. Whichever acid you use, measure it out carefully, and store the acid in a safe place. If you are measuring liquid acid, use a glass or plastic cup, not metal. Add it *slowly* to the water, letting the acid enter at the edge of the water. Rinse the measuring cup in the solution, and stir the mixture with a wooden paddle.

Immerse the pelt in the tanning solution, push it down with the wooden paddle, and stir slowly. Leave the pelt in the solution for five days (or more, up to two weeks if the solution does not get over 75 degrees). Keep the pelt submerged, and stir it gently from time to time.

Neutralize the tanning solution. Remove the pelt and spin out the tanning solution in the "spin" cycle of your washer. Rinse the pelt in clear water twice, then spin out the rinse water. Immerse the pelt in a solution of water and borax, using one ounce borax to each gallon of water. Work the pelt for about an hour in this, then rinse out in clear water. "Spin" out the rinse water. This step is necessary to neutralize the acid solution, so that it does not remain on the skin and damage it.

Tack the pelt out flat, flesh side up. Apply a thin coat of neatsfoot oil to that side. While the oil is soaking in, taking from eight to ten hours in a warm room, you can dry out the wool side, using a fan or hair dryer. Then apply a thin coat of tanning oil or leather dressing on the flesh side.

Drying and softening the pelt. When the tanning oil has soaked in, allow the pelt to dry until it starts showing light colored places. Remove it from the frame, and start the softening process. Stretch the skin in all directions, and, flesh side down, work it over the board, to soften the skin as it finishes drying.

You can sandpaper the flesh side when dry, to make it smooth. Comb out the wool with the coarse teeth of a metal dog comb, and finish with finer teeth. If the wool seems too fuzzy and dried out, you can rub a hair dressing (such as Alberto VO5) on your hands, and rub them lightly through the wool, then brush it gently. Repeat if necessary.

LOCAL SHEARING SERVICES

Learning to shear your own sheep will not only save you the price of hiring a shearer, but it will also give you the convenience of having your sheep sheared when you want it done. It also gains you a skill that can be used to bring in a part-time seasonal income.

In many areas, shearers are scarce and some sheep raisers have to wait until the heat of the summer before they can hire one. If you happen to have only four or six sheep, most professional shearers won't want to spend the time to travel some distance for the small fee that could be charged. Another reason a commercial shearer would not want to do a small number of sheep is that facilities are seldom ideal. Often there is no good method or arrangement for catching the sheep, and no electricity for his shearing equipment.

When you shear with hand shears, which are so convenient for a small number of sheep, you need not worry about electricity. And if you are shearing in your own vicinity you will not travel such a distance, and can have an agreement made ahead that the owner will have the sheep penned when you arrive. With only a few sheep, you can often trade shearing services for the wool from their sheep. They would hardly have enough wool to sell to a wool pool if they were to pay you and keep their wool. In the event they want to keep their wool for their own use in spinning, they will be willing to pay you a fair price for shearing, and expect you to shear the fleece carefully, especially avoiding second-cuts.

When you shear "for the wool," they will want you to trim the hooves and worm the sheep as part of your service, but this can usually be negotiated to include a separate fee. The wool that you get could be sold along with yours to provide income or, if you are a spinner, the best could be selected out for your spinning projects.

CHEESE FROM SHEEP MILK

Another profitable project to consider is gourmet cheese made from sheep milk. Americans import about 20 million pounds of sheep cheese yearly, yet sheep dairying is almost nonexistent in the United States.

The University of Minnesota started a sheep milking research project in 1984, beginning with comparisons of milking abilities of breeds now available here. In their initial tests with purebreds, the Dorset scored highest and adapted well to the milking system, with Suffolk ewes ranking second. In crossbred combinations the following year, Dorset-Lincoln and Rambouillet-Finn ewes produced the most milk. It takes about ten days to train a twelve-month-old ewe, lambing for the first time, to adjust to the handling at milking.

Sheep milk, being high in solids, gives about double the yields of cheese from that of cow milk. Per 100 pounds of milk, sheep milk gives about 20 pounds of cheese, compared to 14 pounds for goat milk and 10 pounds for cow milk.

A common practice among European producers is to permit the lambs to nurse exclusively for thirty days, then wean them and milk the ewes. Weaning at two or three days following birth, with lambs fed artificially, would result in greater total yield of milk. Twice a day milking results in the most milk, but another option is to milk once a day in the morning, followed by lambs sucking. Good nutrition (high protein) is an obvious necessity for high volume of milk.

Taking advantage of sheep milking is *extra* profit. The University of Minnesota gave this example: Suppose you milk 100 ewes, for a total of 20,000 pounds of milk (200 pounds per ewe) worth about $.50 to $.60 per pound. This would total $10,000 to $12,000 from breeding stock you already own and care for and feed, anyway. On a set-up for 100 ewes, it would probably take less than four hours a day to do the milking and clean up, and the season would last from about late April to September. If you want to be done by Labor Day, you just quit milking, for the ewes are slowing down in lactation by then.

On-farm cheese making is possible, but there are always many regulations involving food processing that have to be observed. The factor that makes a cheese-producing cottage industry more possible is that sheep milk, unlike cow milk, can be frozen, then later thawed and made into cheese without losing quality. So, it could be stockpiled, frozen, until there was sufficient quantity for a cheese project.

One way for small sheep operations to go into cheese production would be to form a co-op in order to have a greater volume of milk, and either buy and reactivate one of the small cheese plants, or sell milk to a small cheese factory, or contract with a cheese factory to make the cheese for the sheep raisers to market. The University of Minnesota has been able to market all of their cheese output to gourmet restaurants and food co-ops in the twin-cities area.

MANURE

Another potential income is from the sheep manure, either selling it or using it in your own garden. It not only stimulates the crop growth, but also adds valuable humus to the soil, which is not true of chemical fertilizers. You don't have to be modest about proclaiming its superiority over that of other animals, as the accompanying USDA chart will show:

Pounds per ton of:	Nitrogen	Phosphorus	Potash
Sheep manure	20	9	17
Horse manure	11	6	13
Cow manure	9	6	8

Because sheep make use of ingested sulfur compounds to produce wool, their manure does not have the unpleasant-smelling sulfides found in cow manure. It is also in separate pellets, or in pellets that hold together in a clump, and thus is less messy in the garden, and does not even need aging. If you gather it for your own garden, take it first from paths and places where it does not help to fertilize the pasture. Since it contains many of the valuable elements taken from the soil by the plants eaten by the sheep, it is convenient that they spread a lot of it on the pasture. Its pelleted form causes it to fall in the grass instead of lying on top of it where it might smother the vegetation.

For use in your own garden, clean out the barn twice a year, in spring and fall. The wasted hay and bedding left on the barn floor will have absorbed much of the manure, containing valuable nutrients. Being inside, they are undamaged by rain and sunshine, just waiting to be reclaimed. Spread a thick mulch of this on a portion of the garden and don't even dig it in—just set out tomato, zucchini, and cabbage plants in holes in the mulch, where they will grow without weeding. The portion of the garden which had the mulch this year will have the remains of the mulch completely deteriorated by the following year. By alternating mulched halves, you always have one half heavily mulched for setting out plants, and one half to dig up and plant seeds.

HOMEMADE SOAP

Homemade soap is one of the "good things" of life—and also a profitable small item to add to any product line of sheep-related merchandise. You can make a lot of soap with the fat from lamb or mutton, when it is trimmed for locker packaging. Have the slaughterhouse save all the fat trimmings. Some places will grind them for you, which makes the rendering easier.

Render the tallow. Cut up chunks of lamb or mutton fat (tallow), put it in a large kettle and cook it slowly over low heat. It will take several hours for a large batch, but don't rush or you will risk burning it. When the tallow is all pretty well melted down, strain it through a cloth.

Purify the tallow. Boil the fat that you rendered, with about twice its volume of water. Strain it and set it aside to cool. The clean fat will rise into a solid block. When it has cooled and hardened, remove from the water, turn upside down, cut in wedges, and scrape off the residue of impurities from the bottom. This purified tallow will keep for several weeks in the refrigerator.

Sophia Block's Lamb Tallow Soap Recipe

Measure six pounds of clean purified tallow, which is about 6-3/4 pints of liquid tallow. Heat it slowly in a large enamel pan to between 100 and 110 degrees.

Put 2-1/2 pints (5 cups) of water in a smaller enamel pan. Put the pan on a protected surface. Stand back and slowly pour in one newly opened can of lye (this must be lye, not Drano). Turn your face away so you do not breathe its caustic fumes. The lye will heat up the water. Allow it to cool to 98-100 degrees. Use a candy thermometer, suspended from the side of the pan, not touching the bottom of the pan. When the lye is at the proper temperature, pour it into a half gallon (magnum) liquor bottle, using an agate funnel. Now put the opening of this bottle on the rim of the pot of tallow, and pour the lye mixture very slowly in a thin stream, while stirring the fat and lye together slowly and gently. It is easier if you have a helper to pour in the lye. The tallow should be at the right temperature (100-110 degrees) and the lye poured into it in a very thin stream. Stirring must be done slowly and very gently and steadily. If the lye is poured in too fast, or the stirring is not slow and gentle, the soap will separate or curdle and you will ruin the whole batch. Stir slowly for twenty minutes, and then pour it into prepared containers.

Soap Containers. Agate photo-development pans are ideal for soap. Or use wooden boxes lined with brown paper or with clean cotton cloth, wetted down with water and wrung out. Have the paper or cloth folded out over the outside edge, to make the soap easy to remove when you are ready. You can use cardboard boxes, lined with plastic wrap, which is turned back over the outside edges and stapled to hold it in place while you are pouring the soap.

Pour the soap into these prepared containers, then cover the soap with a board or heavy cardboard, and then with a blanket. This keeps it from cooling too fast. Allow it to cool and harden for a day or two in a warm place away from drafts. The soap will begin to lose its sheen as it hardens. After two or three days and before it gets too hard, you can remove it from the boxes. Cut it into separate bars to age for several weeks, or months, before use. It can be cut neatly with a fine taut wire wrapped around it and pulled tight. Age these bars unwrapped, with air circulating around them, for several weeks. Look for any liquid-appearing substance on or in the soap. That would be free lye, and you should discard the soap or reprocess it.

Soap variations. Mutton tallow soap is often called saddle soap because it cleans and preserves leather so well. It can be used equally well as a bath, laundry, or dishwashing soap, but by a few variations you can make it even more suited to different uses.

Perfumed soap: Add oil of lemon, oil of lavender, or other oil-perfumes (not any containing alcohol) or boil up leaves of rose geranium and use this "tea" as part of the cold water used with the lye. Reserve part of the lye-dissolving quantity of water, boil up the perfuming leaves in it, and add it to the dissolved lye when it has cooled a little. Since soap will absorb odors, it can be perfumed easily after it is in bars and aged, by wrapping it in tissue that has been wet with perfume and dried out.

Green soap: Can be made with "vegetable" coloring obtained by pounding out a few drops of juice from beet tops, or use the vegetable coloring sold for baking.

Mint soap: Use one cup less water to dissolve the lye. Use this cup of water to make a very strong tea from fresh mint leaves. Add this back to the dissolved lye mixture, before adding it to the tallow. Check temperature of lye liquid after adding the mint.

Deodorant soap without chemicals: You can use up to two ounces of vitamin E oil in your soap recipe, adding it to the mixture after stirring in the lye. It has a mild deodorizing quality and is antioxidant, which will prevent any slightly bacony odor if you have used bacon fat along with your tallow.

Honey complexion soap: Add one ounce of honey and stir it slowly into the soap after adding the lye, and before pouring the mixture into the molds.

Laundry soap: To make laundry soap flakes or powder, let the soap age for three or four days. Grate it on a vegetable grater. Dry the flakes slowly in the oven set at "warm," about 150 degrees, stirring occasionally. It can be pulverized when very dry, or just left in flakes.

Dishwashing jelly soap: Shave one pound of hard soap and boil it up slowly with one gallon of water until it is well dissolved. Put it into covered containers. A handful of this will dissolve fast in hot dishwater. For many soft soap and hard soap recipes and variations, see the interesting Garden Way Publishing book, *Making Homemade Soaps & Candles,* by Phyllis Hobson.

MUTTON TALLOW CANDLES

Candles are another good use for the fat that is trimmed off when lamb or mutton is cut and wrapped. While not quite as practical as soap, candles are a fun way to use excess fat and make good gifts, or can be sold.

To prepare candle wicking. Prepared wicking can be purchased (see Sources chapter) but it is simple to make your own from cotton string. One good soaking solution to use is made from eight tablespoons of borax dissolved with four tablespoons of salt in a quart of water. The wicking string is soaked in this for two or three hours, then hung out to dry. Some oldtime candle makers soaked the wicking in apple cider vinegar, or turpentine, and let it dry.

To prepare mutton tallow. Cut up chunks of mutton or lamb fat, put it in a large kettle, and fry it slowly over low heat, as you would for soap. Skim off the bits of fat as they rise to the top. Stir occasionally and do not rush the process and burn the fat. A large batch will take several hours to render out. When it is all pretty well melted down, strain it through a cloth.

Purify it. In a large kettle, dissolve five pounds of alum in ten quarts of water, by simmering. Add the tallow, stir and simmer about an hour, skimming. This not only purifies the tallow, but makes it a little harder texture for use in candles. Cool the tallow until you can touch it comfortably, then strain it through a cloth and set it aside to cool and harden. When it is hard, lift it off the water and scrape off the impure layer on the bottom.

This purified tallow can be stored in a cool place for a week or so until you are ready to make candles, or can be refrigerated or frozen.

Tallow burns with a less pleasant smell than wax or paraffin. It can be perfumed by adding a few drops of pine oil or perfume while the tallow is melted, before dipping or molding the candles.

Candle dipping. Melt the purified tallow and pour it into a wide-mouth jar or container that you can stand in hot water to keep the tallow liquid. Next to this container, have another one filled with very cold water, standing in a pan of crushed ice or ice cubes to keep it cold. Since tallow candles have a tendency to droop in hot weather, don't make your dipped candles too long.

Cut a wick about six inches longer than you want the candle to be, and tie one end of the wick to a small stick. If your containers are large enough, you can tie on several wicks, and dip these all at once.

Dip the wick first into the hot tallow, then withdraw it and let it air-harden for a minute. Then dip it in and out of the ice water, which hardens it. Let it drip thoroughly. Keep repeating this process. To make a tapered candle, do not dip all the way to the top each time you dip it. Since each single dip into the tallow deposits such a thin layer on the candle, it takes a lot of dippings.

Molded candles. It is quicker to mold candles. For candle molds, use plastic or paper cups or cut-down milk cartons. They can be sprayed with a non-stick baking spray (the lecithin-based type) to keep the candles from sticking to the mold, or just brushed with cooking oil. Metal molds should be both oiled and chilled before the tallow is poured in. There are silicone type preparations that are also used for "mold release." As with dipped candles, a shorter and wider shape is best when using tallow, which is not as firm as a wax candle.

Since the bottom of the mold will be the top of the candle, ideally the wick should be threaded out through the bottom, and protrude about an inch. This is easily done when using paper or plastic containers for molds. If you can't make a hole in the bottom of your mold, leave a little coil of extra wick in the bottom that you can pull out when the candle is removed. If you have a wick sticking out the bottom of the mold, you can knot it there so you can pull it straight and tight while pouring in the tallow. It could be fastened at the top to a wire or stick that rests on top of the mold. This would keep the wick straight and centered in the candle until it hardens.

Colored candles. Stir in two teaspoons of powdered household dye, like Rit or Diamond Dye, for each pound of tallow, and mix well into the liquid tallow.

SHEEP CARPENTRY

This book has plans for building various pieces of sheep equipment, and there are plan-services booklets available (see Sources). There is always a need for useful equipment, and you could find a ready sale for duplicates of the pieces you make for your own use.

WASTE WOOL FOR INSULATION

This is a really offbeat use for belly wool and skirtings, the parts you remove before selling your fleeces to spinners. While it may not be legal to use as insulation in residential dwellings, it can insulate barns, sheds, and storage areas quite nicely.

Wash the junk wool well in detergent, rinse it, and squeeze out the rinse water. Treat it with the following solution, suggested by Dr. L.F. Story of the Technical Service of the Wool Organization in New Zealand. It makes the wool both mothproof and fireproof.

> 1 to 2 pounds sodium fluoride (poison—handle with care)
> 4 pounds borax
> 2 pounds boric acid
> 10 gallons of water, mixed with the first three ingredients

Stir the wool in the solution, squeeze out the water, and dry it without rinsing it. Dispose of the surplus fluid where it will not contaminate pastures or water supply. Wash your utensils and your hands well.

When the wool is dry, tease it or card it, and spread it evenly between studs, without leaving air gaps. It may be covered with foil or heavy wrapping paper, if desired.

SPECIAL USES OF WOOL: WOOL IN BEDDING

Wool-filled quilts and comforters are back in style, along with wool-padded mattress-underlays and luxury mattresses with a top layer of wool.

Even the Japanese futon industry is now using wool in much of its traditionally cotton-stuffed bedding. Starting about ten years ago, the amount of wool they used grew to one thousand tons by 1980, then exceeded ten thousand tons in 1987.

Despite its supposed advantage of washability, the polyester-filled quilt is not really satisfactory in use—not warm enough in winter, and too hot in summer. Home quiltmakers are making their quilts washable by making the pieced quilt into a big pillowslip like a duvet cover, removable for washing because the wool batting is quilted in between two cotton muslin sheets. Three of the 6-foot-long batts from a Cottage Industry Carder will fill a standard size quilt.

WOOL SALES TO HANDSPINNERS

The great interest in handspinning has created a specialty market for good fleeces. By keeping the fleeces clean, relatively free of grain, hay, burrs, and other vegetable matter, shearing them carefully (few second-cuts), and handling them properly after shearing, you will have a product that is valuable for handcraft use. The use of sheep coats (see Chapter 17, "Wool and Shearing") can keep the fleeces clean of vegetation.

The fleeces most in demand are the unusual shades of "black sheep" and the fleeces of some of the more exotic breeds, such as Shetland, Icelandic, and Cormo. Even Finnsheep, although not noted for their wool production, have a soft fine wool that is valuable for blending purposes. The real secret of successful selling to handspinners is to offer only the very best of your fleeces, generously skirted. Natural colored dark wool is not as scarce as it was ten years ago, so spinners are becoming much more discriminating as to the quality and condition of fleeces.

MERCHANDISING TO REACH YOUR MARKET

It is useless to have a good product if no one knows about it. There are many ways to find the buyers who want your nice fleeces, and the method you use will be the one most convenient for your situation.

The easiest way is to contact the nearest place teaching spinning and weaving, and leave your name with them or talk to their students. Taking a sampling of fleeces could trigger an immediate response if what you show is of excellent quality.

If you are on a well-traveled road, try putting up a sign saying:

> RAW WOOL AVAILABLE FOR HANDSPINNERS

When you develop a following of spinning customers who appreciate the care you take with your wool, you won't need the sign. They will be regular customers if they are happy with your wool, and will tell their friends.

Try advertising in craft magazines if you are in a remote place with little traffic and few spinners. Good magazines are:

Spin-off
306 N. Washington
Loveland, CO 80537

Shuttle, Spindle and Dyepot
120 Mountain Avenue, B101
Bloomfield, CT 06002

Invitations to an annual shearing day in Wisconsin.

Many sheep raisers put on a real festival for Shearing Day. To do this requires a set day, annually, when you will shear (barring inclement weather, which requires an alternate date). You can see that you must either be able to do your own shearing, or have a very reliable shearer who will promise that date for your use.

I attended a Shearing Day in Michigan a few years ago, and it was more fun than a carnival, with an auction of a prize fleece, shearing and spinning demonstrations, spinning contests, and food booths. The teenage daughter had a booth with *Barbequed Lamburgers*, most delicious and popular. Of course they were mutton, a mixture similar to the *Sloppy Joes* in the "Meat and Muttonburger" chapter, with the addition of a little smoke flavoring.

Another way is to hold such an event after shearing, with all fleeces nicely displayed (skirted, of course). Invite spinners from guilds in the surrounding area, and be sure the event is well publicized.

MONEY IN BLACK FLEECES

Did you know that black lambs have black tongues? The poet Virgil (70-19 B.C.) advised sheep breeders to choose rams lacking pigment in their tongues, if they wanted white fleeces. In many breeds, the dark or part-dark, varicolored tongue indicates a white sheep with recessive dark genes. That sheep could be valuable in a black-sheep breeding program, if its fleece is the kind you want.

The "black sheep" of the sheep family can be the odd dark lamb that crops up occasionally in almost any white breed, the result of recessive genes. In large herds a black sheep is undesirable. Its fleece must be handled and sacked separately. Even in the flock, its black fibers may rub off on the burrs in fleeces of white sheep, or on fences, causing the white wool to be discounted in price because of the special problems caused later in the manufacturing processes.

For handcraft use, the picture is different. Many weavers and knitters are spinning yarn for their own use, and a considerable number are spinning for sale. This has created a special market for fleeces, and the wide variety of colors in natural shades are in demand. As early as 1974, *Shepherd* magazine wrote that "the unwelcome black sheep has suddenly become respectable, with its wool bringing up to several times the price of white wool." But, not just any black fleece—there are so many people now raising dark sheep that it takes a prime fleece, especially clean, to bring top price. There is also the competition from imported fleeces. Australia and New Zealand now have extensive herds of colored sheep, and export tons of their wool to this country.

By far the largest breed listing in *Sheep!* magazine is the column of "Natural Colored" breeders. The same is true in *Shepherd* magazine.

The fleece of most black sheep will tend to lighten, from year to year. In the beginning it may be a disappointment, but in the long run it will prove an advantage, because it gives a greater variation in color from a relatively small flock. So, in shopping for a black lamb, remember that however black she is at birth, she seldom stays jet black, but lightens every year. Don't consider the degree of darkness as the main factor in your selection. Look to her body type and wool grade, which does not change in her lifetime, and probably will be inherited by her offspring. Refresh your memory

Black lambs at the author's home.

about what to look for and what to avoid, in Chapter 1. Be sure you get a healthy animal.

With one or two black sheep to introduce the black genes, you can in a few years work toward the development of a flock of dark sheep, if this is your goal. The easiest way to do this is by the use of a good black ram, of nice wool breed for spinning.

Lambs that will provide spinning fleeces, raised by Wool and Feathers, in Vermont.

If there are spinning classes nearby, ask the teacher what breed of wool is most favored in your area.

Backcrossing for black. Once you get a black ram of a suitable wool type, use him to breed a small flock of white ewes. Geneticists say that the offspring will be "white, but carriers of the black gene." However, in practice we have had people use one of our dark rams on their white ewes and more often than not they got dark lambs.

The first generation of this cross is called the "first filial generation," or F1. If the F1 ewes are bred back to the original black ram, their father, this is called backcross. It produces a generation of F2 lambs, and in theory there should be as many black lambs as white ones, with all carrying the recessive black gene. You can produce quite a flock in a few years by continuing to breed the original F1 ewes to the black ram each year. This much inbreeding is not considered to have a great chance of breeding defects, but it is risky to continue it with succeeding generations (like the F2 offspring). By the time you get a good number of your ram's granddaughters (the black F2 ewes), you would do well to sell the original ram and get a different one, not related to your sheep.

In raising wool for handspinners, the proof of your success will be repeat customers. This depends not as much on the breed of sheep or the wool type as it does on fleece condition. If the wool is poorly sheared, or full of burrs and seeds and other vegetable matter that they must pick out by hand, or if you have not discarded the heavy dung tags, you may sell someone the first fleece, but you're not likely to sell to them a second time. For top prices, you might consider using sheep coats on your sheep (details on this in the chapter on wool).

Hobby spinners enjoy this little hand-turned carder, available with an optional motor unit, for preparing moderate amounts of wool (see Sources).

The selling of dark fleeces alone is not going to support a flock of sheep in the way they would like to become accustomed. But, you can add to this the selling of black sheep breeding stock, and the sale of locker lambs, and have one in your own freezer.

To get an even better price per pound of wool, consider processing it for sale, or handspinning and selling of yarn.

HANDSPUN YARN

Spinning your own wool into yarn is one way to compound its value per ounce. You can spin it for your own use in knitting, or practice on it for home use and go on to sell it to other knitters or weavers when you get good enough. You need to be a knitter in order to successfully sell yarn to other knitters, so you can advise them of the needle size to use, the quantity of yarn they would need, and what size yarn

Avelene McCaul of Indian Springs Farm in Missouri, spinning her fleeces into yarn for sale to knitters and weavers.

to spin for a specific project. The same goes for weaving yarn—unless you can advise your customer, you may be selling them a problem and not know it. There is a list of spinning literature and sources for equipment (see Sources).

WOOL PROCESSING AND CUSTOM CARDING

An article in *New Farm* magazine, April 1984, described one of the wool-processing projects operated by a small sheep farm, Amazing Acres in Pennsylvania. The following is an excerpt from the article which was entitled "20 Times More Income from Wool—With a Cottage-Scale Wool Processing Machine that Paid for Itself in Just Seven Months."

Like all too many wool producers, Donna Kennedy used to sell raw wool to the wholesale market for a mere 47 cents per pound. But now, she's tapping markets that pay $10 per pound and more. The key is a new—affordable—cottage industry wool carding machine that makes Kennedy's wool a real value-added product, increasing its worth at least twentyfold. In addition, she earns extra income with the machine by doing custom processing for other shepherds and spinners.

For Donna Kennedy, the idea of processing her own wool and doing custom work developed as a natural extension of her interest in sheep, wool spinning, weaving, and knitting. She knew that good, clean carded wool was valuable and often hard to get, so why not produce her own and extra to sell. The Cottage Industry Carder, just 48 inches high, 28 inches wide, and 60 to 72 inches long (depending on model), will turn out 6-foot-long batts up to 4 inches thick, for spinning or quilt bats. It can also produce continuous roving, a band of evenly textured wool used for handspinning. The

Extended Cottage Industry Carding Machine with Bump-Winder, produced by Patrick Green Carders Ltd. (see Sources).

new models will produce this roving at a rate of 5000 feet per hour, and automatically wind it into a centerpull "bump" (large ball of carded wool). With its eight carding drums covered with industrial carding teeth, and three sets of steel pressure rolls that compress and draw out the roving before winding, it produces a most elegant product for handspinners. Either natural colored wools or dyed wools can be used, made into solid color or tweedy mixtures, or "rainbow" batts or rovings with several natural or dyed colors layered (see Sources).

HANDCRAFT USE OF UNSPUN FLEECE: AUSTRALIAN LOCKER HOOKING

Australian locker hooking, using unspun fleece, is a new version of the older craft of locker hooking that used commercial rug yarn. The "locker hook" tool has a large crochet hook on one end, and an eye on the other end, so that unspun wool may be hooked into rug canvas, and locked in with a binder yarn carried by the locker hook. This technique offers to nonspinners a way of using their wool in an attractive and profitable manner to create rugs, wall hangings, saddle blankets, and heavy garments. For instruction booklets, see Sources.

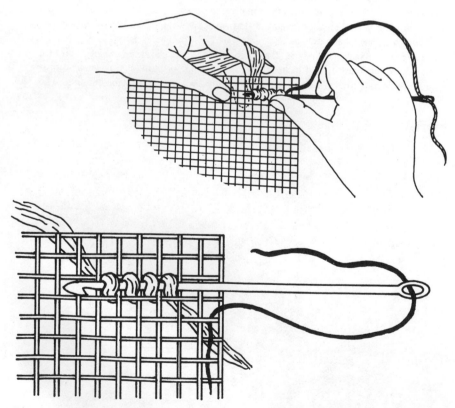

In the locker hooking process, the left hand holds the canvas and the fleece. The right hand does the hooking. For those who are left-handed, the hand positioning may be reversed.

LIVESTOCK DOG BREEDING

There is a good market for well-trained dogs, both herding dogs and sheep guardian dogs. The most valuable guard dogs are those that have already gone through puppyhood (this is the hard part) and are sheep-bonded and ready to go on "active duty." Your experience in their training and use with your own animals will give you a recognizable sales advantage.

INCENTIVE PAYMENTS FOR WOOL SOLD

There is another source of income that is based on your sale of wool, the incentive payment. Incentive payments are founded in the National Wool Act, as amended by the Agricultural Act of 1970 and the Agriculture and Consumer Protection Act of 1973, and are price-support payments not funded directly from tax revenues, but from a tariff charged on lamb and mutton imported into this country. The program was designed to pay the most to the producers of the best quality wool in an attempt to improve the quality of U.S. wool. The county agricultural extension agent can tell you the current status of incentive payments based on your raw wool sales (keep your sales receipts!) as this changes from year to year.

In 1986, the Agricultural Stabilization and Conservation Service started trying to limit the amount that could be paid to the producers of "high value" wools such as those sold to handspinners. This caused the premium wool producers and consumers to band together to form a group called Wool Forum, to protect their interests. For information and address of Wool Forum, see Sources.

MERCHANDISING PRODUCTS

Many enterprising sheep raisers are adding to their income by buying and selling (merchandising) complementary products, such as the Sheep Covers (sheep coats) sold by The River Farm, electric fencing supplies sold by Woodsedge Farms, and particular veterinary products that are otherwise not easily located. Think about what *you* would like to buy, and have had trouble locating—maybe this would be a good product to make available to other sheep people.

TEACHING

Whatever you do well can always be a source of income by teaching the knowledge or skill. A well-organized small sheep operation can offer farm lecture-tours for a fee. "Sheep lore" on-farm classes can be a day class or a weekend bed-and-breakfast offering. Wool handcrafts such as spinning/weaving or locker hooking can be classes or private lessons, all with wool provided. These not only generate income, but create customers for your wool. The only limitation here is your own imagination, inventiveness, and promotion of your availability.

CHAPTER 20

Sheep Guardian Dogs

WHY THE GREAT INTEREST, and recent popularity, of the guardian dog as a protector of sheep? Because sheep owners are becoming desperate, with poisons and traps now barred in so many states.

Coyotes are increasing in number and thrive near civilization, and have become a very real menace to lambs. For small farms in rural areas, however, the greatest threat is from an oversupply of pet dogs. These predators can be more of a danger than coyotes, as one or two dogs can maim and destroy dozens of sheep in a night. One dog attack on a flock can make the difference between a profitable or unprofitable year, and many people have been driven out of the sheep business because of dogs. These are seldom "wild" dogs, or even dog packs—just one or two friendly neighborhood pets, out for a little excitement. Killings occur usually at night or in very early morning, when you are normally asleep. A sheep guard dog is on duty twenty-four hours a day, and most alert and protective during these hours of greatest danger.

Few guard dogs actually *kill* predators, but their presence and behavior reduces or prevents attacks. They may chase a trespassing dog or coyote, but should not chase them far. Chasing for a prolonged distance (or time) would be considered faulty behavior, as the dog should stay near the flock, between the sheep and danger. The best guardians balance aggressiveness with attentiveness to the sheep.

Guardian dogs are most predictably reliable on small farms or ranches with good perimeter fences (preferably electric or electric added to woven wire). It could be suggested that a dog is doubly effective if it is protecting its well-defined property as well as its flock. Good-natured breeds are best for small farms, while more aggressive breeds are needed for large ranches and open range. Long-haired thick-coated breeds are well suited to cold climates, while short-haired dogs do well in the South and West.

The various breeds share similar behaviors. They are ordinarily placid and spend much of the daylight hours dozing. Despite their calm temperament, all of the breeds are fierce when provoked and are wary of intruders, both animal and human. The users of guard dogs say that it would not be possible to place a "dollar value" on the peace of mind that they have, knowing the dog is with their sheep. It can mean greater utilization of pasture, too, since the sheep need not be herded into a night corral.

It takes very few lambs saved by a dog to justify the approximately $300 yearly cost. The cost per year for a dog decreases the longer it lives, as the purchase price is amortized over a longer period. In some situations the mortality rate could be high, from accidents, shootings, or poisoning. You must let all the neighbors know of the dog's presence when you acquire it, so it will not get shot as an intruder. Large signs on perimeter fences can say "Guardian Dog on Duty" so that some well-meaning person will not see the dog among the sheep and shoot it. Other sheep raisers, just passing by, could be alarmed by the sight of the large dog, and take action to eliminate what they believe is a predator.

STARTING WITH A PUPPY

The ideal age to remove a pup from the litter is about eight weeks old, although some claim that dogs placed with sheep before two months old do better than those reared with sheep after two months. Training the dog is primarily raising it with sheep. The process involves supervising to prevent bad habits from developing and to establish limits of acceptable behavior.

The sheep/dog bonding requires training of both the sheep and the dog. Sheep initially accept the puppy easier if they become acquainted in fairly close quarters, but may take a long time to accept the dog if it is turned in with them in a large pasture. The normal procedure is to have a safe enclosure in the sheep area for the new pup, and to let it out with the sheep only when you are present to watch and supervise. A good starting place for a young pup is with a pen of orphan lambs, but if the pup plays too roughly with small lambs, it should be placed with larger lambs that will not tolerate it. Training takes time, persistence and patience, and a pup must be sharply reprimanded for any rowdy behavior with the sheep. The large dog breeds mature slowly and may exhibit puppy behavior up to one and a half years in some cases. A few are reliable with sheep as young as four months, but the average start working at six to nine months old.

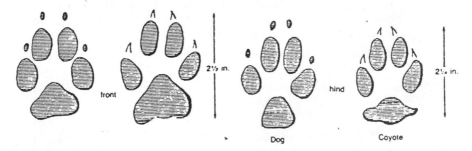

Comparison of dog and coyote tracks. Coyote track is more elongated and not nearly as rounded as a dog track. The front toes on the hind track of the coyote point inward, dog track toes do not.

The primary rule is not to handle, pet, or pick up a puppy—it must bond with the sheep, not humans. This bonding will keep the dog with them and reinforce its protective instinct. Once the sheep and the guard dog have formed a strong relationship, the sheep will seek out the dog, running to it if there is any disturbance.

Oregon State University's Guard Dog Project offers a lot of advice for buying and training a guard dog (see Sources). They give one useful training suggestion to stop a puppy from playing and chasing—the use of a "dangle stick," a board or stick 18 to 30 inches long, which dangles from the swivel hook on the dog's collar. It should hang 3 or 4 inches above the ground when the dog stands upright. This device allows the dog to eat, drink, and display submissive and investigative behavior toward sheep, but when the dog tries to run (chase), the stick gets tangled about his legs. This provides immediate discipline and prevents playful chasing. Use this on a playful pup for three to four weeks, then remove it in stages. First remove the stick but leave the dangle chain, then take away the chain when all playful behavior stops.

It is best not to start with two pups at once, because they will be too inclined to play and may molest the flock in the process. One pup alone will also be more readily bonded to the sheep. Pairing a young dog with an older, experienced dog works better, if you wish to obtain the use of two dogs. The pup is "trained" by the older dog.

Because of the potentially high mortality rate, and the lengthy training needed as the puppy grows up, those who rely on guard dogs as their primary predator control should give thought to having a ready replacement if necessary. Hampshire College's Livestock Dog Project will no longer be breeding dogs, due to liability insurance costs and shifting federal funding, and they suggest that sheep raisers in any given area who depend on guard dogs could jointly maintain a breeding female among the group to raise replacements.

In USDA trials, success rates of guardian dogs did not differ significantly among breeds or between sexes. However, recommendation was made that working dogs be neutered to avoid the problems encountered when the guard dogs or neighboring dogs were in estrus. Neutering is normally done at about four months of age.

The use of bells on some of the sheep is of value, keeping both the owner and the guard dog alerted to disrupted, running sheep.

No amount of proper training and early exposure to sheep will *guarantee* that a dog will become a good guardian. The instinctive ability, strong in the traditional guardian breeds, plays a great part in success. The main attributes needed are:

1. Attentiveness—bonding to sheep and staying with them.
2. Appropriate aggressiveness—growling, barking, fighting if necessary.
3. Defensiveness—staying between the sheep and danger.
4. Trustworthiness—must not harm the sheep.
5. Reliability—wary of unfamiliar humans, but slow to attack.

To select a pup, you need to find a reputable breeder who regularly supplies dogs for sheep guard use. A breeder who raises sheep is probably a good choice. While there is no real way to test the puppy before you raise it, you can observe it in relation to its littermates. Also, observe its parents. They should not be overly shy or aggressive, and should be free from hip displaysia, which is a hereditary joint problem common to large breeds. Pups should have had their shots by eight weeks of age, confirmed by a veterinarian certificate. Most breeders will guarantee the pup for freedom of hip displaysia until at least eighteen months old.

BENEFITS AND PROBLEMS

USDA research at Dubois has indicated that if a dog is properly socialized to sheep from an early age (six to ten weeks) with adequate supervision, there is a high probability that it will successfully guard sheep from predators. They listed the following benefits associated with using a guard dog:

- A reduction in predation.
- Reduced human labor in protecting sheep.
- Better pasture utilization without night corrals.
- Increased acreage use without fear of predators.
- Flock size increase with better use of pastures.
- Improved potential for profit.
- Dog alerts owner to disturbances near the flock.
- Increased self-reliance in predator management.
- Protection for family members and farm property.
- Peace of mind.

Although the majority of dogs that are intended to protect sheep will be successful, there are potential problems, especially during the puppy period. USDA material identified the following potential problems:

- Dog harasses sheep (usually a puppy play behavior).
- Dog does not guard sheep.
- Dog is overly aggressive to people.
- Dog harasses other animals (livestock or wildlife).
- Expenditure of labor to train and supervise dog.
- Dog destroys property (chewing objects, digging).
- Dog is subject to illness, injury, or premature death.
- Dog roams beyond farm boundaries, causing neighbor problems.
- Financial expenditure with no guarantee of dog's success.
- Dog interferes when sheep are moved or interferes with herd dog.

It is unlikely that any one owner will have all of the potential problems or all of the possible benefits of using a dog. For most, the benefit of reduced predation is a great advantage, but for others just a single problem with the animal may be one too many.

RECENT DOG RESEARCH

Although sheep guardian dogs are an ancient tradition in Europe, their use in the United States started in the mid-1970s. A number of sheep and goat producers in Texas turned to dogs in desperation, after other means of predator control had failed.

About the same time, in 1976, Dr. Ray Coppinger founded the Livestock Dog Project at Hampshire College in Amherst, Massachusetts.

The Livestock Dog Project had as its stated goal: to test the effectiveness of livestock guarding dogs as an alternative method of reducing predation. The immediate objective was to introduce, on an experimental basis, guarding dogs to sheep and goat producers. Two dozen dogs of traditional Old World guardian breeds were imported in 1977, the start of the program of training and supplying dogs, for a modest fee, all over the country. All pups between eight and twelve weeks of age were placed on sheep farms, and information on their use and effectiveness in actual sheep flocks was recorded.

The USDA Sheep Experimental Station at Dubois, Idaho, also began studying the economics of guardian dogs in 1978, training and leasing them out, with input required about their use and success. In a questionnaire sent to livestock operators in forty-seven states and Canada who were using their dogs, 90 percent of the 399 responses recommended the use of guard dogs.

Colorado State University at Fort Collins has also been studying and tabulating information about guardian dog behavior, as have the Texas Agricultural Experimental Stations at San Angelo and Sonora. Oregon State University has done research and testing and puts out pamphlets of useful information on the subject (see Sources).

GUARDIAN DOG BREEDS

GREAT PYRENEES

The Great Pyrenees is a native of the Pyrenees Mountains between Spain and France, and is said to have a common ancestry with the Saint Bernard. They are mostly pure white, with a rough coat, and a most impressive size with weights from 100 to 125 pounds.

This is the gentlest of the guard dogs, probably because they have been bred in the United States as pets for many generations. In USDA trials at Dubois, Idaho, this was the only one of the breeds that did not at any time bite a human. Although they are, in most instances, further separated from their traditional guardian ancestry, they have proven reliable when raised and bonded to sheep at an early age, and are the most common of the guardian breeds in the United States.

KOMONDOR

The Komondor (plural is Komondorok) means "corded coat," and it has a tremendous coat of hair that hangs in locks similar to that of an Angora goat. This may require some maintenance to take out burrs. It is considered to be of Hungarian breeding, but may have been brought there by Turkish Kun families, migrating with their sheep and dogs in the thirteenth century.

This dog has a very serious disposition, is a devoted guard, wary of strangers, and an independent thinker. While used in Hungary to protect herds, it was also used

Great Pyrenees with sheep at Ambrosia Farm, Chepachet, RI.

Adult Komondor at USDA Sheep Experiment Station, Dubois, ID.

for protecting property and factories. USDA tests found it more successful with pastured sheep than with sheep on open rangeland. This breed has been bred for 1,000 years to operate with a great deal of independence, but that independence must be carefully channeled by a firm and loving master, for the dog to be effective.

MAREMMA (Maremmano Abruzzese)

This guardian dog has a sleepy-eyed, relaxed look, with a rough coat, usually white. The Maremmas have been used in the Italian mountains to guard sheep for centuries, usually two or three per flock to protect all sides from wolves. They do well in teamwork protectiveness. In Italy their ears are usually docked as pups, to prevent a wolf from getting a grip on the head.

This is an independent-minded dog, but will obey single commands taught as a puppy. They still interpret commands in terms of context and duty—loyalty to the flock always prevails. It is one of the most successful breeds used in the Livestock Guard Dog Program, and is know to be one of the calmest of the guardian breeds during the daytime, but with instinctive nocturnal alertness.

Maremma pup meets lamb at Hampshire College's Livestock Dog Project.

KUVASZ (plural Kuvaszok)

This dog has a rough white coat, dark lips, eyes, skin, and nails. The males weigh 100 to 130 pounds, the females 90 to 110 pounds. It is probably a native of Hungary and was used there for many years. Many Kuvasz were killed there during World War II, sadly depleting the original stock.

The Kuvasz is independent and not easily obedience trained—"no" must be strictly enforced. It is very protective of its own property, and once having learned the boundaries will protect them fiercely. The females seem more alert, while the males are more apt to kill predators. They are able and agile runners and catch or

Maremma guard dogs at Woodsedge Wools, Stockton, NJ. Adult, left. 5 month old pup, right.

corner a predator easily. While capable of functioning without supervision (after proper training), this breed seems to have an emotional need for a certain amount of human company.

Kuvasz with lamb, at Lala Kingsbury's Spinning Wheel Sheep Farm, Frankfort, ME.

Shar on migration to summer pasture in Yugoslavia. Photo by Ray Coppinger, Hampshire College.

A Shar puppy exhibiting proper "sheep attentiveness." (Pat Devore's New Horizons Sheep Farm, Peace Valley, MO)

SHAR PLANINETZ (Sarplaninac)

The first Shar known to be imported into the United States (1975) was carried down the mountains of Macedonia in a basket on the back of a donkey. This breed has been used traditionally by shepherds of Macedonia and reported to have been a court guardian of kings. Its name comes from the mountain range of Macedonia in southeastern Europe, with the area of Yugoslavia as its origin. While similar to the Maremma or Pyrenees, it is a bit smaller, occasionally white, but usually tan to dark, and often black.

This dog has a quiet gentle temperament, and many have been trained and distributed by the Livestock Dog Project at Hampshire College.

ANATOLIAN SHEPHERD (Karabash = black head, Akbash = white head or tricolor).

This breed originates in Turkey, known there as "Coban Kopegi"—shepherd dog. It is one of the more aggressive of the guardian breeds, even to human strangers, but can be trained by socialization to be friendly toward visitors if this is desirable. Any strangers must be introduced to the dog, with the owner making sure the dog accepts them. It is very possessive toward family, property, and livestock.

To obtain the most effective guarding, it is best to avoid too much training, so as not to interfere with its instinct and independent intelligence. Necessary commands would be "come," "stay," "no," and necessary training would be to walk on a leash when necessary, and to allow being handled.

Male Anatolian Shepherd with sheep on Toni Tooker's ranch in Somerset, CO.

Four-year-old Tibetan Mastiff, very patient with his sheep and his puppies on farm of Michael Morgan, Tonasket, WA.

TIBETAN MASTIFF

This is one of the oldest breeds in existence today, and its lack of genetic problems is evidence of centuries of natural selection and survival of the fittest. It is black with tan markings, including tan spots over the eyes, has a distinctive double coat with a ruff around the neck and shoulders, and carries its full tail well over its back. In its native land, these dogs travel with caravans of the Tibetan sheepherders and traders, defending the herds and the tents of their masters from such predators as wolves, snow leopards, and robbers. They are loyal to master and flock, with antipathy to strangers. Bitches have a single annual estrus, approximately every ten to twelve months (the mark of a primitive breed) and a lack of dog odor.

Publications

SHEEP RAISING

Guide for Lambing Problems, $5
from:
  R.F. Cross
  1655 Linwood Drive
  Wooster, OH 44591

Computer program for scheduling sheep care, based on desired lambing date
from:
  Norseman Sheep Company
  Route 2, Box 141A
  Wellsville, KS 66092

Breeds of Sheep, $4.50 postpaid
from:
  Jones Sheep Farm
  Route 2, Box 185
  Peabody, KS 66866
(booklet with pictures and info on 25 breeds of sheep)

Sheepman's Production Handbook, $20
Control of Foot Rot in Sheep, no charge
Pelts Make a Profit Difference, $.50
Pack It with Pride—Shearing, $.50
all from:
  Sheep Industry Development Program
    Inc.
  200 Clayton Street
  Denver, CO 80206

WOOL INCENTIVE PROGRAM LOBBY

"Wool Forum," for information on incentive program regulation, and how we can protect premium wools
from:
  Wool Forum
  Route 1, Box 153
  Hanning, MN 56551-9740

Wool Handbook, 1-LD SCOAP, no charge
from:
  USDA/ASCS
  Management Services Division
  P.O. Box 2415
  Washington, DC 20013
This is the handbook the county offices use. When you order, tell them you are a producer.

GUARD DOGS

National Yugoslavian Sarphaninac Livestock Guardian Dog Association
  New Horizons Sheep Farm
  Route 1, Box 30
  Peace Valley, MO 65788

MAGAZINES

Sheep! Magazine
  Box 329
  Jefferson, WI 53549
(the best)

Black Sheep Newsletter
  Route 1, Box 288
  Scappose, OR 97056

Southern Sheep Producer
  Route 2
  Arlington, KY 42021
(has Dr. Schaffer's column)

Shepherd Magazine
  5696 Johnson Road
  New Washington, OH 44854

National Wool Grower
  600 Crandall Building
  10 West Stars Avenue
  Salt Lake City, UT 84101
(useful, but for large sheep raisers)

Sheep Breeder and Sheepman Magazine
P.O. Box 796
Columbia, MO 65205
(mostly for large sheep raisers)

The Southern Sheepman
P.O Box 350
Loganville, GA 30249

Sheep Canada Magazine
Box 113
Moosehor, Manitoba ROC 2E0
CANADA

SID Research Digest
Sheep Industry Development Program
  Inc.
200 Clayton Street
Denver, CO 80206

SPINNING AND WEAVING MAGAZINES

Spin-Off
Interweave Press
306 N. Washington Avenue
Loveland, CO 80537

Handwoven
Interweave Press
306 N. Washington Avenue
Loveland, CO 80537

Shuttle, Spindle and Dyepot
120 Mountain Avenue, B101
Bloomfield, CT 06002

Northwest Woolgather's Quarterly
2058 14th Avenue
Seattle, WA 98119
(mostly northwestern states info)

The Heddle
P.O. Box 1600
Bracebridge, Ontario POB 1C0
CANADA

SHEEP GUARD DOGS

Introducing Livestock-Guardian Dogs
Pamphlet #EC1224, $.50
and
Raising and Training a Livestock-Guardian Dog
Pamphlet #EC1238, $.50
both from:
  Agricultural Communications
  Publication Orders
  Administration Building, Room 422
  Oregon State University
  Corvallis, OR 97331-2119

Guardian Dogs Protect Sheep from Predators, $.25 from:
  Agricultural Research Service
  U.S. Sheep Experiment Station
  Dubois, ID 83423

Training a Stock Dog, for Beginners, $8
postpaid from:
  R.J. Karrasch
  Route 1, Box 42
  Vandiver, AL 35176

The Pearsall Guide to Successful Dog Training
by Margaret E. Pearsall, published by
  Howell Book House, Inc.
  230 Park Avenue
  New York, NY 10169

The Dog Owners Home Veterinary Handbook
by Carlson and Giffin, published by
  Howell Book House, Inc. Address above.

Informational pamphlets available
from:
  New England Farm Center
  Livestock Dog Project
  Amherst, MA 01002
  (Hampshire College)

BOOKS ON SPINNING, CRAFTS, COTTAGE INDUSTRY

Spinning and Weaving with Wool
by Paula Simmons, $15.95 postpaid
from:
  Paula Simmons
  48793 Chilliwack Lake Road
  Sardis, BC V2R 2P1
  CANADA

Spinning for Softness and Speed
by Paula Simmons, $8.60 postpaid
from Paula Simmons. Address above.

Handspinner's Guide to Selling
by Paula Simmons, $10.50 postpaid
from Paula Simmons. Address above.

Turning Wool into a Cottage Industry
by Paula Simmons, $9.95 postpaid
from Paula Simmons. Address above.

Australian Locker Hooking by Joan Z. Rough
Booklet and locker hook $6.95 plus postage of $1.35 from:
  Fox Hollow Fibers
  560 Milford Road,
  Earlysville, VA 22936

Australian Locker Hooking by Signe Nickerson
Booklet $2.50 postpaid
Locker Hook $3.25 postpaid from:
  The Crafty Ewe
  Box 161
  Danville, WA 99121

Catalog of craft books, $1 from:
  The Unicorn
  1304 Scott Street
  Petaluma, CA 94952

CHEESE MAKING

Practical Sheep Dairying from:
  New England Cheesemaking Supply
  P.O. Box 85
  Ashfield, MA 01130

Making American-Type Cheese at Home
FS266, $.25
and
Making Soft Cheeses at Home
FS227, $.25
both available from:
  Agricultural Communications
  Publication Orders
  Administration Building, Room 422
  Oregon State University
  Corvallis, OR 97331-2119

Farmstead Cheese, 217 pages, $5 from:
  Dr. E. A. Zottola
  Department Food Science & Nutrition
  University of Minnesota
  St. Paul, MN 55108

Making Cheese at Home, AG Bulletin
0504, 23 pages, $1 from:
  Distribution Center
  University of Minnesota
  3 Coffey Hall
  St. Paul, MN 55108

Profitable Milk Production from Sheep,
from:
  U.S. Feed Grains Council
  1400 K Street N.W., Suite 1200
  Washington, DC 20005

Sheep Milking Equipment, from:
  Gascoigne Milking Equipment, Ltd.
  Edison Road, Houndsmills
  Basingstoke, Hampshire
  ENGLAND RG21 2YJ

ELECTRIC FENCING

Building an Electric Antipredator Fence
PNW 225, $.75 from:
    Agricultural Communications
    Publication Orders
    Administration Building, Room 422
    Oregon State University
    Corvallis, OR 97331-2119

The New Fencing Systems Made Simple,
$4 from:
    Premier
    Box 89S
    Washington, IA 52353

Reprints of 1987 Power Fence articles
from Southwestern Sheepman Magazine,
$2 from:
    Sonny Reeves
    929 Folds Road
    Carrollton, GA 30117

POISON PLANTS

*Protecting Dairy Goats from Poisonous
Plants*
WREP 46, $.50 from:
    Agricultural Communications
    Publication Orders
    Administration Building, Room 422
    Oregon State University
    Corvallis, OR 97331-2119

BUILDING PLANS

Plans for Farm Buildings and Equipment,
$2.50 from:
    Extension Service
    · Utah State University
    Logan, UT 84321
(catalog showing many plans for which
they can supply drawings)

*Sheep Handbook of Housing and Equip-
ment*, $6.00 from:
    Midwest Plan Service
    Iowa State University
    122 Davidson Hall
    Ames, IA 50011
(detailed working drawings for sheep
equipment)

TANNING

Home Tanning Methods
Leaflet # 21005,
By G.M. Spurlock, free from:
    Co-Op Extension Service
    University of California
    Davis, CA 95616

INFORMATION ON CHARGERS

    Parker-McCrory Manufacturing Co.
    2000 Forest Avenue
    Kansas City, MO 64108

SHEEP SUPPLIES

Sheepman Supply Co.
P.O. Box 100
Barboursville, VA 22923
$.50 catalog

Hand shears, electric shears, ear tags, lamb nipples, worming supplies, fence supplies, medications. Mighty Mike Sheep Elastrators and bands, Footvax, hoof shears, tanning kits, Prolapse retainers, mineral mixes, Covexin-8 and other vaccines, bells, zinc sulfate, automatic waterers, etc.

Nasco Farm and Ranch
901 Janesville Avenue
Fort Atkinson, WI 53538

Ear tags, barbed wire dollies, capsule forceps, sheep coats, fence stretchers, ear tags, shepherd's crooks, etc.

C.H. Dana Co., Inc.
Hyde Park, VT 05655

Sheep bells, sheep coats and blankets, hay hook, barbed wire stretchers, etc.

Sabian Corporation
5301 N. 1st Street
P.O. Box 2647
Abilene, TX 79604

Coyote snare formerly available from Clyde Thate.

Dakota Stockman's Supply
Box 12
Brookings, SD 57006

Veterinary thermometer, KRS maggot-killer bomb, wound-kote bomb, etc.

Mid-State's Livestock Supplies
125 East 10th Avenue
South Hutchinson, KS 67505

Newborn Lamb Carrier and other sheep supplies. Sunbeam electric sheep shears.

Premier Sheep Supplies
Box 89
Washington, IA 52353

Lambing rope (like soft veterinary fingertips for lambing), new Prolapse Harness, and other sheep supplies including Felco #2 shears for hoof trimming.

Uckele Health Equipment
P.O. Box 160
Blissfield, MI 49228

TM salt w/selenium, Bovatec, Ivomec, Levasole wormer, Naselgen, Covenix-8, Vibrio, and other vaccines. Selenium-E, disposable needles and syringes.

Peterson's Natures Way
Route 2-2584
Sidney, MT 59270

Lamb adoption coats, Colostrum powder, Colostrum sheep boluses.

Norseman Sheep Co.
Route 2, Box 141A
Wellsville, KS 66092

Computer program for sheep, newborn lamb carrier, send long SASE for information.

Northbend Meadows
Route 1, Box 110
Dalbo, MN 55017

Sheep Covers (sheep coats), send SASE for information.

Coal Creek Sheep & Wool Co.
7542 Coal Creek Drive
Superior, CO 80027

Sheep Covers (sheep coats), send SASE for information.

Protein Technology
701 4th Avenue, South
Suite 1350
Minneapolis, MN 55415

"Colostryx®," milk whey antibody product for lambs.

ELECTRIC FENCE SUPPLIES

Premier Fence Systems
P.O. Box 89
Washington, IA 52353

Snell Gallagher
Permanent High Tensile Fencing
from:
Sonny Reeves
929 Folds Road
Carrollton, GA 30117
and other dealers

Parker-McCrory Manufacturing Co.
Parmak Fence Chargers
2000 Forest Avenue
Kansas City, MO 64108

TANNING .

Theo Stern Tanning Co.
334 Broadway
Box 55
Sheboygan Falls, WI 53085

L & M Fur and Woolen Enterprises, Inc.
Erie & Belmont Avenues
Quakertown, PA 18951

Bucks County Fur Products
P.O. Box 204
Quakertown, PA 18951

Rittel's Tanning Supplies
170 Dean St.
Taunton, MA 02780

COTTAGE INDUSTRY WOOL EQUIPMENT

Wool carders and pickers, all sizes
from:
Patrick Green Carders Ltd.
48793 Chilliwack Lake Road
Sardis, BC, CANADA V2R 2P1

SHEEP MILKING EQUIPMENT AND CHEESEMAKING EQUIPMENT

Alfa-Level Inc.
11100 N. Congress Avenue
Kansas City, MO 64153
Attn: Mr. Scott Sanford
National Product Manager, Milking

Westfalia Systemat
1862 Brummel Drive
Elk Grove, IL 60007
Attn: Mr. Warren Dorathy
Sales Manager

New England Cheesemaking Supply Co.
P.O. Box 85
Ashfield, MA 01130

SAUSAGE-MAKING SUPPLIERS

Norseman Sausage Supplies
Route 2, Box 141A
Wellsville, KS 66092
Send SASE for catalog.

SHEEP ASSOCIATIONS

BLACK SHEEP
Natural Colored Wool Growers Association
18150 Wild Flower Drive
Penn Valley, CA 95946

BORDER LEICESTER
American Border Leicester Association
Beverly Tiffany, Secretary
7594 SR 534
West Farmington, OH 44491

CHEVIOT
American Cheviot Sheep Society
Ruth Bowles, Secretary
RR 1, Box 100
Clarks Hill, IN 47930

American North Country Cheviot Sheep
Association
Ann Trimble, Secretary
717 Fall Creek Road
Longview, WA 98632

CLUN FOREST
North American Clun Forest Association
E.F. Reedy, Secretary
High Meadow Farm
Ferryville, WI 54628

CALIFORNIA RED SHEEP REGISTRY
Paulette Soulier
Route 2, Box 2140
Davis, CA 95616

COLUMBIA
Columbia Sheep Breeders Association
Richard L. Gerber, Secretary
P.O. Box 272E
Upper Sandusky, OH 43351

COOPWORTH
Coopworth Sheep Association
Pauline and Ernie Copp
HCR 85, Box 336
Bonners Ferry, ID 83805

CORMO
American Cormo Sheep Association
P.O. Box 549
Monticello, UT 84535-0549

CORRIEDALE
American Corriedale Association
Russel Jackson, Secretary
Seneca, IL 61360

COTSWOLD
American Cotswold Record Association
Registrar, Elaine Vogt
18150 Wild Flower Drive
Penn Valley, CA 95946

DELAINE MERINO
American Delaine Merino Association
Elaine A. Clouser, Secretary
1193 Township Road, 346
Nova, OH 44859

DORSET
Continental Dorset Club
Mrs. Marion A. Meno, Secretary
P.O. Box 506
Hudson, IA 50643

FINNSHEEP
Finnsheep Breeders Association
Mrs. Claire Carter, Secretary
P.O. Box 512
Zionsville, IN 46077-0512

HAMPSHIRE
American Hampshire Sheep Association
P.O. Box 345
Ashland, MO 65010

JACOB
Jacob Sheep Society
242 Ringwood Road, St. Leonards
Ringwald, Hampshire
ENGLAND 8H24 2SB

KARAKUL
American Karakul Sheep Registry
Julie O'Neill, Secretary
Route 1, Box 179
Rice, WA 99167

LINCOLN
National Lincoln Breeders Association
Teresa M. Kruse, Secretary
R.R. 6, Box 24
Decatur, IL 62521

MINOR BREEDS
American Minor Breeds Conservancy
Box 477
Pittsboro, NC 27312

MONTADALE
Montadale Sheep Breeders Association
Mildred Brown, Secretary
P.O. Box 44300
Indianapolis, IN 46244

OXFORD
American Oxford Sheep Association
Wally W. Watts, Secretary
R.R. 4
Ottawa, IL 61350

PERENDALE
American Perendale Breeders Association
Norlaine Schultz
1811 New Hampshire Avenue
Ashton, MO 20702

POLYPAY
American Polypay Sheep Association
Linda Wick
Route 2, Box 2172
Sidney, MT 59270

Polypay Sheep Association of America
Pauline Holmes, Secretary
3618 Green Acre Drive, Jacks Valley
Carson City, NV 89701

RAMBOUILLET
American Rambouillet Breeders Association
LaVerne McDonald, Secretary
2709 Sherwood Way
San Angelo, TX 76901

ROMELDALE
Romeldale CVM (California Variegated
Mutant) Registry
Route 1, Box 265-A
Rossville, GA 30741

ROMNEY
American Romney Breeders Association
John H. Landers, Jr., Secretary
4375 N.E. Weslinn Drive
Corvallis, OR 93771

SCOTTISH BLACKFACE
Scottish Blackface Sheep Breeders Association
Richard Harward, Secretary
39282 River Drive
Lebanon, OR 93755

SHROPSHIRE
American Shropshire Registry
Elizabeth Glasgow, Secretary
P.O. Box 1970
Monticello, IL 61856

SOUTHDOWN
American Southdown Breeders Association
Florence Strouse, Secretary
Route 4, Box 148
Bellefonte, PA 16823

SUFFOLK
American Suffolk Sheep Society
VeNeal Jenkins, Secretary
1115 N. Main Street
Logan, UT 84321

National Suffolk Sheep Association
Kathy Krafka
P.O. Box 324N
Columbia, MO 65205

TARGHEE
U.S. Targhee Sheep Association
Vicky Arnold, Secretary
Box 40
Absarokee, MT 59001

TUNIS
National Tunis Sheep Registry
Leona Fitzpatrick, Secretary
Route 1
Wayland, NY 14572

Other Garden Way Publishing Books You Will Enjoy

Raising Milk Goats the Modern Way, by Jerry Belanger. Complete up-to-date coverage by the editor of *Countryside & Small Stock Journal*. Fencing, breeding, kidding, goat milk products and much more. 160 pages, 6x9, illustrated quality paperback, $6.95. Order #062-4.

Raising Rabbits the Modern Way, by Bob Bennett. Revised and Updated Edition. The best book available on modern rabbit raising techniques, with complete coverage of building wire hutches, feeding, breeding, marketing and all aspects of managing small rabbitry. 192 pages, 6x9, illustrated quality paperback, $8.95. Order #479-4.

Raising Poultry the Modern Way, by Leonard Mercia. Stock selection, feeding, brooding, management, disease prevention for laying flocks, meat chickens, turkeys, ducks & geese, written by an Extension Service expert especially for the small grower. 224 pages, 6x9, illustrated quality paperback, $8.95. Order #058-6.

Raising the Home Duck Flock, by Dave Holderread. For persons who want to raise ducks but don't know how to get started. Explains how many ducks you should have for various conditions, which breeds may be best for you, where and how to buy them. 192 pages, 6x9, illustrated quality paperback, $7.95. Order #169-8.

Small-Scale Pig Raising, by Dirk van Loon. From the beginning, this book explains exactly *why* raising a feeder pig is the best bet for someone with little land who wants to produce the most meat for the smallest investment of time and money. 272 pages, 6x9, illustrated quality paperback, $10.95. Order #136-1.

Keeping Livestock Healthy, by N. Bruce Haynes, D.V.M. Updated Edition. A how-to-do-it veterinary guide. Covers all farm animals. "A must." — American Veterinary Journal. 324 pages, 6x9, illustrated quality paperback, $14.95. Order #409-3.

Tan Your Hide! Home Tanning Leathers & Furs, by Phyllis Hobson. Easy methods for the home tanner and fur skin worker. 134 pages, 6x9, illustrated quality paperback, $6.95. Order #101-9.

Basic Butchering of Livestock & Game, by John J. Mettler, Jr., D.V.M. With this easy-to-read text you will find: 140 how-to drawings (very easy to follow every step); recommended tools and equipment; processing and preserving; meat inspection information. 206 pages, 6x9, illustrated quality paperback, $10.95.   Order #391-7.

Raising a Calf for Beef, by Phyllis Hobson. All the information you need to raise a calf, with complete butchering instructions. 128 pages, 6x9, illustrated quality paperback, $6.95.   Order #095-0.

The Canning, Freezing, Curing & Smoking of Meat, Fish & Game, by Wilbur F. Eastman, Jr. An authoritative work on the art of home processing of meat, fish and game. Step-by-step instruction. 220 pages, 6x9, illustrated quality paperback, $7.95.   Order #045-4.

Building Small Barns, Sheds & Shelters, by Monte Burch. Provides basic, easy-to-follow construction methods for attractive outbuildings. Plans for multi-purpose barns and barn-style garages, woodshed, toolshed, carport, and housing for poultry, rabbits, and hogs. 236 pages, 6x9, illustrated quality paperback, $12.95.   Order #245-7.

These books are available at your bookstore, feedstore, lawn and garden center, or directly from Garden Way Publishing, Dept. 8800, Schoolhouse Road, Pownal, VT 05261. Send for our free mail order catalog. **Please send $2.00 per order for postage and handling.**

INDEX